Elements of Synchrotron Light

Elements of Synchrotron Light

for Biology, Chemistry, and Medical Research

Giorgio Margaritondo

Faculté des sciences de base
Ecole Polytechnique Fédérale de Lausanne
Switzerland

OXFORD
UNIVERSITY PRESS

*This book has been printed digitally and produced in a standard specification
in order to ensure its continuing availability*

1007160700

OXFORD
UNIVERSITY PRESS

Great Clarendon Street, Oxford OX2 6DP

Oxford University Press is a department of the University of Oxford.
It furthers the University's objective of excellence in research, scholarship,
and education by publishing worldwide in

Oxford New York

Auckland Cape Town Dar es Salaam Hong Kong Karachi
Kuala Lumpur Madrid Melbourne Mexico City Nairobi
New Delhi Shanghai Taipei Toronto
With offices in
Argentina Austria Brazil Chile Czech Republic France Greece
Guatemala Hungary Italy Japan South Korea Poland Portugal
Singapore Switzerland Thailand Turkey Ukraine Vietnam

Oxford is a registered trade mark of Oxford University Press
in the UK and in certain other countries

Published in the United States
by Oxford University Press Inc., New York

ISBN 978-0-19-850931-8

Printed and bound by CPI Antony Rowe, Eastbourne

To my brother
Professor Enrico Margaritondo, MD

Preface

Laudato sie, mi' Signore cum tucte le Tue creature,
spetialmente messor lo frate Sole,
lo qual è iorno, et allumini noi per lui.
Et ellu è bellu e radiante cum grande splendore:
de Te, Altissimo, porta significatione.

Praised be, my Lord, with all Your creatures,
especially our brother the Sun,
which is the light of the day, and by which You illuminate us.
It is beautiful and radiating with high brightness:
It brings an image of You, the Highest.

St Francis, Laudes Creaturarum

The light of the righteous shines brightly,
but the lamp of the wicked is snuffed out.

Proverbs 13:9

The objective of this book is to present a simple, practical and broad picture of synchrotron light sources and of the corresponding experimental techniques. The book is specifically written for beginning or prospective synchrotron users in chemistry, the life sciences, medical research and related fields.

The domain broadly called 'synchrotron radiation' was initiated in the late 1960s as a fallout of elementary particle physics. The experts running particle accelerators had known for decades that electrons circulating at high speed in an accelerator ring emit electromagnetic waves—*synchrotron light*—over a very wide spectrum ranging from infrared radiation to X-rays. They considered this to be a problem, since the emission causes the loss of part of the electron energy necessary for elementary particle experiments.

A different view, however, had already emerged in the 1950s: the emitted waves had unique and potentially very useful properties. Sources of visible and infrared light had made spectacular advances culminating with the development of lasers. By contrast, ultraviolet radiation and X-ray sources were much less advanced: progress had been marginal since Röntgen's discovery of X-rays. Synchrotron light promised a very effective solution to this problem.

The practical difficulties, however, were very serious. The main applications of ultraviolet light and X-rays were in condensed matter physics, a discipline mostly confined to small laboratories and involving small research groups. Using large facilities implied serious logistic problems and the need to break entrenched cultural barriers. Furthermore, the practical use of synchrotron light was still largely untested: synchrotron-based research was considered—with good reasons—a risky idea.

This situation changed with the first successes of synchrotron-based research in the late 1960s and early 1970s. A strong expansion began that still continues more than three decades later. Instead of using accelerators built for other disciplines, synchrotron-based research could increasingly rely on machines specifically developed to emit and use synchrotron light. This became a multi-billion-dollar worldwide enterprise.

During the expansion, physics ceased to be the leading area of synchrotron research: the domain expanded to chemistry and materials science. In recent years, a new change has occurred: the life sciences and medical research have increased their presence and, according to most projections, are becoming the leading areas.

This is not surprising: ultraviolet light and X-rays are excellent tools to investigate chemical bonds and related properties, including the atomic structure of solids and molecules. On the other hand, the life sciences increasingly seek a microscopic understanding of biological phenomena with a particular focus on molecular structures. Superior-quality X-ray sources are, therefore, crucial instruments for the future of this broad research area.

The evolving spectrum of disciplines brings many new users in contact with synchrotron facilities. This positive tendency is attended by a serious problem: the new users' insufficient knowledge of synchrotron light and of the related techniques. In most cases they are not interested in the source properties as long as it produces for them photons with the required characteristics. This is a rather damaging attitude: by better understanding the sources, users are often able to invent new techniques and improve their research.

The problem is aggravated by the communication barrier between synchrotron experts and new users—who are confronted with formidable obstacles when they try to penetrate the misteries of synchrotron light. Even introductory presentations are typically based on the complicated formalism of advanced electrodynamics.

In 1985, as a response to this problem, I wrote a book entitled *Introduction to synchrotron radiation*, primarily for users from physics. The choice was reasonable since physicists and a few chemists constituted almost the totality of the user population at that time. The book was quite well received and provided several generations of researchers with a simplified introduction to this blooming field.

That text, however, is of limited use for medical and biological scientists. The explosive expansion of synchrotron use in these domains stimulated me to write the present book. The objectives are to further simplify the 1985 presentation by limiting the formalism to a bare minimum, and also to update many of the concepts presented 17 years ago. My ideal readers are beginning graduate students: should my work help them in their career, I will consider this effort to be a complete and rewarding success.

I am very grateful to the many colleagues who helped me in writing this book with their suggestions. I am particularly indebted to Fabia Gozzo and Lenny Rivkin for their critical reading of parts of the manuscript and for their excellent suggestions.

G. M.

Lausanne, Switzerland
May 2002

Contents

User's guide

This book is primarily written for users who do not possess a broad background in physics and chemistry but only basic notions. Its main objective is to explain in simple terms the properties of synchrotron sources, the related instrumentation and the utilization techniques of specific interest to chemistry, biology and medical research. Readers from other disciplines could also profit from the simple presentation, primarily based on qualitative concepts rather than on complicated mathematics or physics.

In order to facilitate understanding, the discussion is presented at two different levels: a main body plus specialized insets. The main body is self-contained and can be used as a first reading or even as the only reading for an audience not interested in mathematics and physics details. The insets contain more detailed discussions of some important notions presented in the main body.

Even the insets are kept at a very simple mathematical level, so they should be broadly comprehensible. Since each inset is independent of the others, the reader can select a subset of topics for deeper analysis, thereby tailoring the reading to his/her personal needs and interests.

1. Smart-tourist guide to a synchrotron light facility

A person visiting a synchrotron laboratory for the first time is confronted with a rather confusing picture: a huge room cluttered with metal tubes, aluminum foil and mysterious electronic equipment, where scientists and technicians are engaged in a variety of tasks. Gradually, a coherent structure becomes evident (Fig. 1.1): in the center of everything, there is a closed-loop tube. Several straight tubes depart from it in tangential directions; each one passes through complicated devices and ends with a formidable-looking piece of equipment. This is where most of the scientists and technicians are gathered.

This general structure reflects the basic operation of the facility. The central closed-loop tube is a 'storage ring' into which electrons are sent by a special device (the 'injector') and where they circulate for many hours at very high speed and energy. Left alone, the electrons would move along straight trajectories. What keeps them circulating within the storage ring is a series of magnets, which bend and control their trajectories.

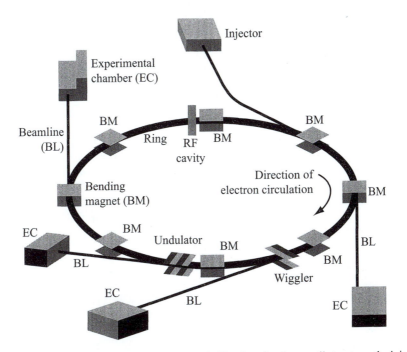

Fig. 1.1 The many components of a synchrotron facility fit a simple overall structure: the injector sends electrons into the ring where they circulate for hours under the effect of bending magnets (BM) and other magnetic devices. The bending magnets, as well as the 'undulators' and the 'wigglers' force the electrons to emit synchrotron light that is collected by the beamlines (BL) and brought to experimental chambers (EC) for a wide variety of applications.

3

Fig. 1.2 A real synchrotron facility: aerial picture of Elettra in Trieste, Italy. A synchrotron source is used by hundreds or even thousands of scientists, and its storage ring has a circumference of hundreds of meters .

Such magnets play two roles in fact: while bending the trajectory of the electrons, they also force them to *emit electromagnetic waves*. This phenomenon is improperly called the emission of 'synchrotron light' because it was first detected in the electron accelerators known indeed as 'synchrotrons'. Nowadays, the typical synchrotron sources are accelerators of the type called 'storage rings'.

The emission and use of synchrotron light is the scope of the facility. The tangential straight tubes are 'beamlines', which collect the emitted synchrotron light, improve its characteristics and channel it into experimental chambers—the formidable-looking pieces of equipment surrounded by busy people.

Synchrotron light thus becomes available for a wide variety of applications: to analyze the physical and chemical properties of materials and biosystems, to identify their structure down to the level of atoms and also to fabricate novel microscopic devices. In the early days of synchrotron research, almost all applications were confined to physics and chemistry. This picture has radically changed, and the 'busy people'—the synchrotron light users—now include not only physicists and chemists but also biologists, medical researchers, engineers, environmentalists and materials scientists.

What stimulates this multidisciplinary interest, thus justifying the efforts to build and operate these huge facilities? In order to answer this, we must explain in general why synchrotron light is so important for science and technology.

1.1. Electromagnetic waves as a research tool

Electromagnetic waves are a basic tool used to explore the physical and biological world. As new-born babies we begin to learn by 'seeing' things with light, which consists of electromagnetic waves. As we grow up and become more sophisticated, we can use different types of electromagnetic waves to explore different properties of the world around us: for example, infrared light to study atomic-level vibrations or X-rays to study the atomic structure of molecules. Scientific progress is, to a large extent, based on building better sources of electromagnetic waves and using them as effectively as possible.

A synchrotron facility is an excellent source of the most useful electromagnetic waves to explore materials and biological systems: X-rays and ultraviolet light (and, to some extent, infrared light). To fully understand this point, we must discuss two issues. What specifically makes the waves emitted by a synchrotron source so powerful? Why do we need such a complicated facility to produce them?

1.1.1. The most important properties of a synchrotron source

We need different types of electromagnetic waves for different research applications. The difference is primarily based on the property known as 'wavelength'. This property is easy to understand if we think of visible sunlight. Sunlight has no special color: it is 'white'. On the other hand, Isaac Newton demonstrated with his famous prism experiment that white light is really a superposition of different colors.

Single-color light is a propagating electromagnetic wave consisting of a pure oscillation. Imagine (Fig. 1.3a) a snapshot of such a wave at a given time: we see oscillations with a regular periodicity. The distance λ over which the wave repeats itself is the wavelength.

Imagine now the wave traveling at the speed of light, $c \approx 3 \times 10^8$ m/s. With a detector placed at a fixed site, we detect an oscillating signal as a function of time. A full oscillation corresponds to the arrival at the detector of one wavelength; this takes a total time $T = \lambda/c$, which is called the 'period' of the wave.

During a time interval Δt, we detect $\Delta t/T$ oscillations. Thus, the number of detected oscillations per unit time is $v = 1/T = c/\lambda$; this quantity is called the 'frequency' of the wave. Wavelength and frequency are thus linked by the following simple equation:

$$v = \frac{c}{\lambda}. \tag{1.1}$$

When selecting the right type of waves for a given application, we must identify what wavelength is needed. For example, many biological properties are determined by the spatial positions of atoms in macromolecules. The technique used to measure such positions is crystallography, which is based on the diffraction of electromagnetic waves by atoms.

Crystallography requires wavelengths from a few angstroms to a fraction of an ångström [one ångström (Å) equals 10^{-10} m]. The corresponding frequencies (eqn 1.1), are in the 10^{17}–10^{19} s^{-1} range. This wavelength range or 'spectral range' corresponds to the waves known as 'X-rays'.

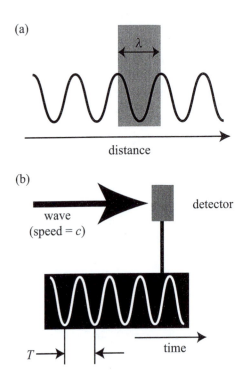

Fig. 1.3 Definition of the wavelength (λ) and period (T) of an electromagnetic wave.

Consider now the use of electromagnetic waves to explore chemical properties, which are also very important for materials and biological systems. The waves interact with the system under investigation, and we can extract the desired information by exploiting our knowledge of the interaction mechanisms—for example, absorption or scattering.

When absorbed by a material, light behaves as a collection of small chunks of energy called 'photons'. Specifically, the total absorbed energy equals an integer number of photons. Each photon has energy E given by the famous Albert Einstein's equation:

$$E = h\nu,\qquad(1.2)$$

where $h \approx 6.6 \times 10^{-34}$ joule is the Planck constant.

To explore chemical properties, the photon energy must not be too far from the typical energy of a chemical bond. This is of the order of a few electronvolts, equivalent to a few times 10^{-19} J. Equation 1.2 would then predict $\nu = E/h \approx 10^{15}$ s^{-1}. The corresponding wavelength $\lambda = c/\nu$ (eqn 1.1) is of the order of 10^{-7} m or one thousand ångströms. Waves with such a wavelength belong to ultraviolet light.

We thus reach a fundamental conclusion: to explore chemical bonds in solids, molecules and biosystems, we need X-rays and ultraviolet light. In turn, chemical bonds and the related atomic-level structures determine most of the important properties of materials and life-science systems. Thus, advanced research absolutely requires good sources of X-rays and of ultraviolet light.

For visible light, we can count on a wide variety of sources: fire, sunlight, torchlight, incandescent lamps, lasers, etc. Unfortunately, nature is not similarly benevolent when supplying ultraviolet light and X-rays. Natural sources are quite bad, and artificial sources were rather poor until the advent of synchrotrons in the late 1960s. Synchrotron sources eliminated the problem by supplying very high-quality beams of X-rays and ultraviolet light, many orders of magnitude 'brighter' (see the next section) than previous sources.

1.1.2. How is synchrotron light generated?
We now analyze the mechanism enabling synchrotron sources to produce electromagnetic waves and in particular X-rays and ultraviolet light. The starting point is a discussion of what makes a light source 'good', i.e. useful for practical applications. We certainly like to have a lot of light: thus, a 'good' source must emit a large amount of energy per unit time—a large power.

1.1.2.1. The source 'brightness'
High power, however, is not enough. Compare an ordinary lamp with a laser pointer used by a lecturer (Fig. 1.4). The power of the lamp is higher, but the laser is more effective and more useful because it concentrates the emitted power precisely where we need it.

Similarly, for most research applications we must concentrate as much wave power as possible into a small specimen area. This is much easier if the source itself possesses a high *brightness* (or *brilliance*)—which means that it emits a lot of light from a small source area, and also that the emission occurs within a narrow angular cone.

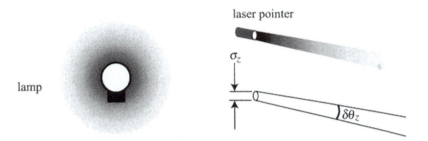

Fig. 1.4 Use of the parameter *brightness* to define the quality of a source: a lamp (left) may emit more power than a laser pointer (top right), but the laser is a better-quality source because its emitting area is more concentrated and its emission is more angularly collimated. The bottom right part shows in the vertical (z-axis) direction the parameters σ_z (source size) and $\delta\theta_z$ (angular spread) used in the definition of brightness, eqn 1.3; the other two parameters σ_y and $\delta\theta_y$ are the source size and the angular spread in the horizontal direction.

The source brightness is, by definition, proportional to the emitted power F and inversely proportional to the source sizes in the horizontal (y) and vertical (z) directions, $\sigma_y \sigma_z$, and to the angular spreads in the same direction, $\delta\theta_y \delta\theta_z$:

$$\text{brightness} = \text{constant} \times \frac{F}{\sigma_y \sigma_z \delta\theta_y \delta\theta_z}. \tag{1.3}$$

Equation 1.3 implies that the brightness can be increased by increasing the emitted power and/or by decreasing the source size and angular spread. Consider now the specific case of X-rays: high brightness is desirable, but very difficult to achieve. A conventional X-ray source for radiology does not match the requirements for high brightness: the emitting area is large, and the emission occurs in a broad range of directions.

Furthermore, the bad geometry of a conventional X-ray source cannot be easily corrected by optical devices. In the case of visible light, a divergent beam can be concentrated into a small area with a simple and rather inexpensive focusing lens. There are no similar lenses for X-rays, which either pass almost undisturbed through glasses or are absorbed by them.

X-ray focusing lenses can be based on the reflection of X-rays—which is quite weak— or their diffraction. Such devices are complicated, delicate and rather expensive.

A good source geometry is therefore much more important for X-rays than for visible light. The lack of good sources with high brightness severely limited the effectiveness of X-rays for over eighty years, from their discovery to the advent of synchrotron facilities.

1.1.2.2. Relativity at work

How does a synchrotron source produce bright X-rays? The key is a clever use of Einstein's relativity. Before discussing this point, we must see how electromagnetic waves are created in general.

An electromagnetic wave is a perturbation of the electromagnetic field which, after being created at a given site, propagates away from it at the speed of light. This is similar to the waves caused by a stone hitting the surface of a lake. The stone produces a local perturbation in the water level. Such a perturbation cannot remain localized: it propagates on the lake surface as a wave.

The emission of electromagnetic waves thus requires a local perturbation of the electromagnetic field. A radio antenna provides a good example: when no voltage is applied, there is no current in the antenna and no emission of waves. When an oscillating (AC) voltage is applied, an oscillating current is created which causes a local perturbation of the electromagnetic field that propagates as electromagnetic waves.

Without moving, an electric charge in the antenna cannot emit electromagnetic waves since its electric field is constant and without perturbations. Charges moving at constant speed create a current with a constant, non-perturbed magnetic field: once again, there is no wave emission. To emit electromagnetic waves we need *accelerated* electric charges. The oscillating charges in the radio antenna emit waves because their motion is not at constant speed, but accelerated.

What is the frequency of the wave emitted by the antenna? Not surprisingly, it corresponds to the oscillation frequency of the charges. Could we then produce X-rays using the same strategy? Could we build suitable 'radio antennas' to emit them?

The answer is negative because of the magnitude of the corresponding frequencies. Radio waves emitted by an ordinary antenna have typical frequencies in the kilohertz or megahertz range (10^3–10^6 s^{-1}). We have seen that X-ray wavelengths correspond to much higher frequencies. Building an antenna and an oscillating electronic circuit with that kind of frequency is impossible.

Imagine, however, a slightly different strategy. Rather than forcing the electrons to oscillate in an antenna, we shoot them towards a periodic array of magnets, as shown in Fig. 1.5. Each individual magnet applies a 'Lorentz force' to the moving electrons, slightly deviating their direction. The periodic magnet array thus forces the electrons to gently 'undulate' around a straight line (the electron trajectory without magnets).

Such a magnet array is called an *undulator* and is the basic X-ray emitting source in a modern synchrotron facility. Seen from the front end of the undulator (Fig. 1.5), each moving electron looks like an oscillating charge in a radio antenna, it thus emits electromagnetic waves. This approach overcomes the limitations of real radio antennas and can produce very high frequencies corresponding (eqn 1.1) to very low wavelengths.

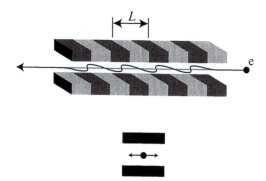

Fig. 1.5 Top: schematic explanation of an *undulator* source: a periodic array of magnets (with period L) causes small undulations of an electron (e) that would otherwise move along a straight line at high speed $\approx c$. Bottom: seen from the end (the beamline) the electron looks like an oscillating charge in a radio antenna: it thus emits electromagnetic waves.

How can this occur? Essentially, by a combination of three factors:

- First, the electrons are sent into the undulator at very high speed—almost the speed of light. An electron completes one full undulation in the time period L/c required to travel along $L =$ one magnet array period. Thus, the emitted frequency is given by $v = c/L$ and the emitted wavelength (eqn 1.1) is $\lambda = L$. This, however, does not automatically lead to the production of X-rays, since we cannot fabricate a magnet array with a period L as short as the X-ray wavelengths.

- Fortunately, the period L and the corresponding wavelength λ are automatically shortened by a relativistic effect. Seen from the point of view of a fast-moving electron, the length L shrinks because of the so-called 'Lorentz contraction' (see Inset A)— and so does λ.

- Furthermore, the emitted wavelength is detected in the reference frame of the beamline, which is different from the reference frame of the electron. In the beamline or laboratory frame, the wave source—the electron—moves with speed close to c. Its emission is thus subject to the phenomenon known as the 'Doppler effect'—like the changing pitch of a train whistle as it moves towards us (see again Inset A). The Doppler shift further contracts the undulator-emitted wavelength, producing the desired X-ray values.

In summary, a synchrotron source emits short-wavelength X-rays by using a macroscopic-sized object such as a periodic magnet array and then by shortening the emitted wavelength with the combination of two relativistic effects: Lorentz contraction and Doppler shift. In a synchrotron facility there are other emitting devices besides undulators—such as wigglers and bending magnets. All of them exploit a combination of relativistic effects, like the undulators.

Inset A: Relativistic background of synchrotron light

Relativity is the key to understanding how a synchrotron source works. For our discussion, we need a simple introduction to the basic concepts of relativity and an equally simple discussion of the two relativistic phenomena used by synchrotron sources: Lorentz contraction and Doppler shift.

The experimental foundation of relativity is a rather surprising phenomenon: the speed of light, c, does not change when it is measured in two different reference frames, one moving at constant speed u with respect to the other. This fact is surprising because for objects like cars or trains the speed does change with the reference frame.

However surprising, the invariance of c is not a conjecture but a solid fact, supported by many experiments. What are its consequences?

Consider an electron moving at speed u along the x-axis, and emitting synchrotron light. Imagine an emitted light pulse along the x-axis; during a time period Δt_e, the pulse travels along a distance Δx_e such that

$$\frac{\Delta x_e}{\Delta t_e} = c . \tag{A1}$$

Δx_e and Δt_e are the values measured from the point of view (reference frame) of the moving electron. If we look at the light pulse from the beamline reference frame (the 'laboratory frame'), then the measured position and the measured time change from x_e and t_e to x_L and t_L.

Before Einstein's relativity, this change was believed to follow the simple rules:

$$x_L = x_e + ut_e \tag{A2}$$

$$t_L = t_e . \tag{A3}$$

These rules, however, are in conflict with the experimental fact that c does not change with the reference frame. In fact, according to eqns A2 and A3 the speed of the light pulse in the new frame would be

$$\frac{\Delta x_L}{\Delta t_L} = \frac{\Delta x_e + u\Delta t_e}{\Delta t_e} = \frac{\Delta x_e}{\Delta t_e} + u = c + u .$$

Thus, c would not be invariant but change from c to $c + u$, contrary to all experimental evidence. We can eliminate this problem by adopting a modified version of eqns A2 and A3:

$$x_L = \gamma(x_e + ut_e), \tag{A4}$$

$$t_L = \gamma(t_e + \alpha x_e) , \tag{A5}$$

where γ and α are two parameters to be determined. How? First of all, we must require the invariance of c. In the laboratory frame:

$$\frac{\Delta x_L}{\Delta t_L} = \frac{\Delta[\gamma(x_e + ut_e)]}{\Delta[\gamma(t_e + \alpha x_e)]} = \frac{\Delta x_e + u\Delta t_e}{\Delta t_e + \alpha\Delta t_e} = \frac{\dfrac{\Delta x_e}{\Delta t_e} + u}{1 + \alpha\dfrac{\Delta x_e}{\Delta t_e}} .$$

If $\Delta x_e/\Delta t_e = c$ (speed of light), then $\Delta x_L/\Delta t_L$ must be also equal to c. Thus:

$$\frac{c + u}{1 + \alpha c} = c ,$$

and $c + u = c + \alpha c^2$, which gives $\alpha = 1/c^2$.

Equations A4 and A5 can then be written as:

$$x_L = \gamma(x_e + ut_e), \tag{A6}$$

$$t_L = \gamma\left(t_e + \frac{ux_L}{c^2}\right). \tag{A7}$$

Note a remarkable implication of eqn A7: a time interval Δt_e measured from the point of view of the electron (at the position $x_e = 0$) changes to:

$$\Delta t_L = \gamma \Delta t_e \tag{A8}$$

when measured in the laboratory frame. Thus, since $\gamma > 1$ (see later), the time interval *appears expanded*.

We must now evaluate the parameter γ in eqns A6 and A7. Imagine (see Fig. A-1) a light pulse that, instead of traveling along the x-axis, follows a direction y perpendicular to x. From the point of view of the electron, after a time Δt_e the light pulse reaches the distance:

$$y_e = c\,\Delta t_e . \tag{A9}$$

Seen from the laboratory (see again Fig. A-1), the pulse travels in an oblique direction. After a time $\Delta t_L = \gamma \Delta t_e$, it reaches the distance y_L along the y-direction and the distance $x_L = u\Delta t_L$ along the x-direction. Thus, the total distance traveled $c\,\Delta t_L$ is equal to $\sqrt{y_L{}^2 + (u\Delta t_L)^2}$, so $c^2\Delta t_L{}^2 = y_L{}^2 + u^2\Delta t_L{}^2$, and

$$y_L = \sqrt{c^2 - u^2}\,\Delta t_L\,;$$

considering eqn A8, we obtain:

$$y_L = \sqrt{c^2 - u^2}\,\gamma \Delta t_e . \tag{A10}$$

On the other hand, the y-coordinate, being perpendicular to the motion, cannot change when we change the reference frame:

$$y_e = y_L , \tag{A11}$$

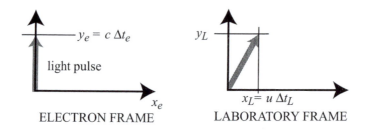

ELECTRON FRAME   LABORATORY FRAME

Fig. A-1

which, using the results of eqns A9 and A10 for y_e and y_L gives

$$c = \sqrt{c^2 - u^2}\,\gamma,$$

or:

$$\gamma = \frac{1}{\sqrt{1 - \dfrac{u^2}{c^2}}}. \tag{A12}$$

The Doppler shift

We now possess all the ingredients to calculate the wavelength and frequency changes of synchrotron light when we change our point of view from the electron frame to the laboratory frame. Imagine (Fig. A-2) a moving source emitting a series of ultrashort pulses at a time distance T_e from each other. The frequency in the source reference frame is $\nu_e = 1/T_e$, and the corresponding wavelength is $\lambda_e = c/\nu_e = cT_e$.

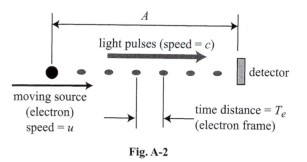

Fig. A-2

We can evaluate the frequency in the laboratory frame from the measured time distance between two successive pulses. Assume that the first pulse is emitted at the time $t_e = t_L = 0$ s and detected by a detector at distance A: the first-pulse detection occurs A/c s after the emission. The second pulse is emitted after the time period T_e, which expands (eqn A8) to γT_e in the laboratory frame. In the meantime, the source motion has shortened the distance A to $A - u(\gamma T_e)$. Thus, the second pulse is detected at the time:

$$\gamma T_e + \frac{A - u\gamma T}{c}.$$

The interval between the two detection times is:

$$T_L = \gamma T_e + \frac{A - u\gamma T_e}{c} - \frac{A}{c} = \gamma T_e \left(1 - \frac{u}{c}\right).$$

Thus, the frequency in the laboratory frame is

$$v_L = \frac{1}{T_L} = \frac{T_e}{\left(1 - \frac{u}{c}\right)} = \frac{v_e}{\gamma\left(1 - \frac{u}{c}\right)}. \tag{A13}$$

Equation A13 shows the Doppler shift in frequency between the source (electron) frame and the laboratory frame. The corresponding Doppler shift between the wavelengths $\lambda_e = c/v_e$ and $\lambda_L = c/v_L$ is

$$\lambda_L = \lambda_e \gamma\left(1 - \frac{u}{c}\right). \tag{A14}$$

In a synchrotron, the electron speed u is $\approx c$, and therefore $u/c \approx 1$. Thus* we have $\gamma(1 - u/c) \approx 1/(2\gamma)$, and eqn A14 becomes:

$$\lambda_L \approx \frac{\lambda_e}{2\gamma}, \tag{A15}$$

which is the approximate Doppler shift that will be used to derive eqn 1.4.

The Lorentz contraction
Why does the undulator period L appear contracted from the point of view of the moving electron? To answer, consider a modern way to measure L: interferometry. One could for example (Fig. A-3) use mirrors to mark the distance L, and adjust λ_L to obtain constructive interference between the initial wave and the wave that has traveled back and forth along L.

The condition for constructive interference at the semi-transparent mirror is:

$$\frac{L}{\lambda_L} + \frac{L}{\lambda_L} = n,$$

$$(*) \quad \gamma\left(1 - \frac{u}{c}\right) = \frac{1 - \frac{u}{c}}{\sqrt{1 - \frac{u^2}{c^2}}} = \frac{\sqrt{1 - \frac{u}{c}}}{\sqrt{1 + \frac{u}{c}}} = \frac{\sqrt{1 - \frac{u^2}{c^2}}}{1 + \frac{u}{c}} = \frac{1}{\gamma\left(1 + \frac{u}{c}\right)} \approx \frac{1}{2\gamma}.$$

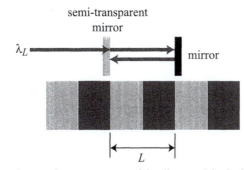

Fig. A-3 Interferometric measurement of the distance L in the laboratory frame.

where n = an integer number; thus:

$$n = \frac{2L}{\lambda_L}.$$ (A16)

Imagine now repeating the interferometric measurement in the electron frame, calling L' the result. When the wave travels back from the mirror, its wavelength is Doppler shifted (eqn A14) to $\lambda_L \, \gamma(1 + u/c)$. During the path to the mirror, the Doppler shift is given by the corresponding expression with reversed speed direction, $\lambda_L \, \gamma(1 - u/c)$.

Thus, the constructive-interference condition for the entire wave path becomes:

$$\frac{L'}{\gamma\left(1 - \dfrac{u}{c}\right)\lambda_L} + \frac{L'}{\gamma\left(1 - \dfrac{u}{c}\right)\lambda_L} = n = \frac{2L}{\lambda_L}.$$

After a few steps, this equation gives

$$L' = \frac{L}{\gamma}$$ (A17)

confirming the Lorentz contraction from L to L/γ that will be used to derive eqn 1.4.

The electron energy

Einstein's best known law says that the electron energy is m_0c^2, where the electron (rest) mass m_0 is $\approx 9 \times 10^{-31}$ kg. The quantitative value is $m_0c^2 \approx 9 \times 10^{-31} \times (3 \times 10^8)^2 \approx 8 \times 10^{-14}$ J. If we measure the energy in electronvolts [1 electronvolt (eV) $= 1.6 \times 10^{-19}$ J], then m_0c^2 is approximately equal to 0.5×10^6 eV, or 0.5 MeV [1 MeV equals one million electronvolts].

The value m_0c^2 is the energy of the electron measured in the *electron* reference frame—the so-called 'rest energy'. In the laboratory frame, this value changes because the electron mass changes from m_0 to γm_0. Thus:

$$\text{electron energy} = \gamma m_0 c^2. \tag{A-18}$$

Why this change? We can give a plausibility argument. In the non-relativistic small-speed limit ($u/c \ll 1$), the difference between the laboratory-frame energy and the rest energy must be the classical kinetic energy $m_0 u^2/2$; this implies:

$$m_0 c^2 \approx \frac{m_0 u^2}{2}, \text{ thus } \gamma \approx 1 + \frac{u^2}{2c^2},$$

and this is indeed true: in fact, if $u/c \ll 1$, then:

$$\gamma = \frac{1}{\sqrt{1 - \dfrac{u^2}{c^2}}} \approx \frac{1}{1 - \dfrac{u^2}{2c^2}} \approx 1 + \frac{u^2}{2c^2}.$$

Equation A-18 shows that the γ-parameter corresponds to *the ratio between the electron energy* (in the laboratory frame) *and the electron rest energy* ≈ 0.5 MeV. For example, a ring of energy 1 GeV = 1000 MeV gives $\gamma \approx 2000$.

In what follows we will often find mathematical equations including the γ-parameter. We must always keep in mind that the γ-parameter corresponds to the electron energy and therefore changes with it.

1.1.2.3. Undulator emission

We can now expand our analysis of undulators a little. The key parameter to describe the relativistic properties of moving electrons (Inset A) is the 'gamma-parameter' (γ), proportional to the electron energy.

This energy can be measured in joules, but the most commonly used unit in the synchrotron business is the GeV, equivalent to one billion electronvolts or 1.69×10^{-10} J. An energy of 1 GeV corresponds (Inset A) to $\gamma \approx 2000$. For synchrotron sources the energy ranges between 0.25 and 8 GeV, and γ between 500 and 16,000.

The γ-factor determines (see again Inset A) both the Lorentz contraction and the Doppler shift—and therefore the emitted wavelength of an undulator. The Lorentz contraction shrinks L to L/γ. This is also the emitted wavelength in the reference frame of the electron. Observed from the laboratory, this wavelength is Doppler-shifted by a factor $\approx 1/(2\gamma)$, becoming:

$$\lambda_L \approx \frac{L}{2\gamma^2}. \tag{1.4}$$

Consider a typical undulator period $L = 5$ cm $= 0.05$ m: a wavelength of that magnitude would correspond to radio waves. If the electron energy is 1 GeV and $\gamma \approx 2000$, then eqn 1.4 predicts $\lambda_L \approx 6 \times 10^{-9}$ m or ≈ 60 Å, which is in the X-ray range.

We can express the same result in terms of photon energies rather than of wavelengths. Equation 1.2 shows that the photon energy is directly proportional to the frequency ν and therefore (eqn 1.1) inversely proportional to λ:

$$E = h\nu = \frac{hc}{\lambda}. \tag{1.5}$$

Commonly used units are the ångström (Å) for λ and the electronvolt (eV) for the photon energy. The corresponding practical version of eqn 1.5 using these units is

$$E[\text{eV}] \approx \frac{1.2 \times 10^4}{\lambda[\text{Å}]}. \tag{1.6}$$

For example, a wavelength of 60 Å corresponds to a photon energy ≈ 200 eV.

1.1.2.4. A laser-like angular collimation

Undulators are very effective at producing X-rays, since the electrons move 'freely' and this limits energy losses. Thus, they meet the first condition for a bright source: they can emit a lot of power. In addition, the brightness (eqn 1.3) is enhanced because this power is emitted from a small area $(\sigma_y \sigma_z)$ and with small angular spread $(\delta\theta_y, \delta\theta_z)$.

As far as the source area is concerned, a single electron in a straight trajectory would be an extremely small point-like source. The reality is a bit different: an electron in an undulator does not move in a straight line but along a slightly undulating trajectory.

Furthermore, each time the electron completes a circuit around the ring and reaches the undulator its conditions are slightly different, and so is the trajectory. Finally, in the ring there is not one single electron, but many different electrons moving along slightly different trajectories. All these factors increase the effective size of the undulator source. Even so, the source remains exceedingly small, as required for high brightness.

What about the angular spread? Is it sufficiently small? The answer is an unqualified yes: the undulator emission challenges the angular collimation of a laser, even if the emission mechanisms are very different. The extreme collimation of the undulator emission is specifically caused by the forward motion of the emitting electrons.

To understand this point, consider the sound waves of a whistling train (Fig. 1.6). When the train is standing in a station the waves are emitted with roughly equal intensity in all directions. If the train moves, its velocity is combined with the wave motion in the forward direction: the sound is 'projected forward' becoming collimated.

Similarly, the electromagnetic waves emitted by a moving electron in an undulator are 'projected forward'. This phenomenon, because of relativistic effects, is much more pronounced than for sound waves. The resulting angular spread is determined by the ubiquitous γ-factor (see Inset B), its value in radians being $\approx 1/\gamma$.

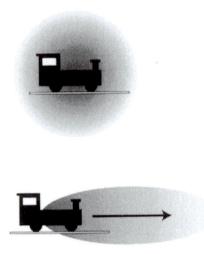

Fig. 1.6 Collimation of sound waves due to the motion of the source. Top: a standing whistling train emits waves over a broad range of directions. Bottom: if the train moves, the sound is projected forward and becomes angularly collimated. A similar phenomenon occurs for the electromagnetic waves emitted by an electron in an undulator. However, the relativistic speed of the electron strongly enhances the resulting angular collimation.

Thus, for a source of 1 GeV the γ-value of ≈ 2000 corresponds to an angular spread of approximately one-half of a milliradian. This is a remarkably small value that contributes to very high brightness. We will see, however, that the 'natural' angular spread $\approx 1/\gamma$ is not the only factor determining the overall angular spread of the emitted beam (Section 1.3.1). For an undulator, this implies both good news and bad news.

Inset B: Extreme angular collimation—another relativistic effect

How does relativity produce the extreme angular collimation of the undulator emission (and of synchrotron light in general)? The answer is quite simple.

Consider the emitting electron in its own reference frame. The angular emission pattern has the typical geometry of radio waves produced by an antenna: it thus occurs in a very broad angular range.

Consider now one specific light beam in the xy plane (Fig. B-1). In the electron frame, the emission angle θ_{ye} is determined by the components c_{xe} and c_{ye} of the beam velocity:

$$\sin\left(\theta_e\right) = \frac{c_{ye}}{c_{xe}}. \tag{B1}$$

Since $c_{ye} = \Delta y_e/\Delta t_e$ and $c_{xe} = \Delta x_e/\Delta t_e$, the ratio c_{ye}/c_{xe} equals $\Delta y_e/\Delta x_e$.

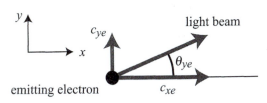

Fig. B-1 Emission geometry of a light beam in the xy plane (electron reference frame).

Consider now the corresponding emission angle θ_{yL} in the laboratory frame, given by the ratio $\Delta y_L / \Delta x_L$. Equations A-6 and A-11 imply that by transforming Δx_e into Δx_L we introduce a γ-factor into the denominator of $\Delta y_L / \Delta x_L$—whereas $\Delta y_e = \Delta y_L$. Thus, $\Delta y_L / \Delta x_L$ is proportional to $1/\gamma$. Since $1/\gamma$ is quite small, we have

$$\theta_{yL} \approx \sin\left(\theta_{yL}\right) = \frac{\Delta y_L}{\Delta x_L} \approx \frac{1}{\gamma}.$$

All emission angles in the electron frame are thus compressed to $\approx 1/\gamma$ in the laboratory frame, and the angular spread along y is

$$\delta\theta_y \approx \frac{1}{\gamma}. \tag{B2}$$

A similar conclusion holds for the z-direction and for the corresponding angular spread $\delta\theta_z$. This, however, is only the 'intrinsic' angular spread caused by the emission mechanism. The actual angular spread is also affected by the electron undulations and by other factors—but after all corrections it still remains very small.

1.2. A 'real' synchrotron source

We must still learn additional things about undulators. First, however, we can take a well-deserved pause. Our guide brings us into the real world, explaining for example how an undulator fits into the overall picture of a synchrotron facility (Fig. 1.1).

The core element of the facility is the electron storage ring. This is the tube under ultrahigh vacuum where electrons circulate for many hours. 'Ultrahigh' vacuum means a pressure lower than 10^{-13} atmospheres. This vacuum is required to limit the loss of electrons due to scattering by residual atoms in the atmosphere.

The 'ring' tube includes straight sections and curved sections. On several straight sections undulators are inserted. The emission of each undulator is collected by a beamline departing from the ring vacuum tube. Optical devices (described in the next chapter) process the emitted waves to improve their characteristics.

Then, the waves reach the experimental chamber at the end of the beamline, which includes the necessary equipment for the specific use of synchrotron light. The scientists we noticed during our visit are the users of the experimental chambers.

Besides the undulators, other devices are used around the ring to produce electromagnetic waves—notably, wigglers and bending magnets. With many emitting devices feeding light into different beamlines, a synchrotron can simultaneously support many different experiments. The construction and operating costs are thus partitioned over a large number of research groups and applications, alleviating the financial impact for each of them.

A real electron storage ring is of course more complicated than a simple tube. The electrons must be produced, pre-accelerated and injected into the ring: this is done by adevice called an 'injector', which can be a synchrotron or a linear accelerator (LINAC).

After injection, the electrons must be kept circulating in the ring. Their trajectories are periodically deflected by the bending magnets—which are big dipole electromagnets placed at curved sections of the ring. The bending action is based on the Lorentz force, the same force we found in undulators. This force is active whenever a moving electric charge is subject to a magnetic field. Its magnitude is proportional to the charge, to its speed and to the field strength, and its direction is perpendicular to the field and to the charge velocity. A force perpendicular to the velocity does not produce any work, so it cannot change the energy nor the speed of the electron. But it does change the *direction* of the velocity, bending the electron trajectory.

Bending magnets are complemented by other magnets, together forming the 'magnet lattice' of the storage ring. For example, quadrupole magnets improve the geometrical characteristics of the electron beam. One of their objectives is to reduce the transverse section of the beam and its angular spread, which contribute to the effective synchrotron source size and angular spread and influence the brightness (eqn 1.3).

The circulating electrons move slightly differently from each other, and this increases the cross-sectional area of the electron beam. This area is the actual size of the synchrotron source that determines its brightness (eqn 1.3). Furthermore, deviations of the electron path from its ideal trajectory contribute to the angular spread of the emission, increasing it beyond its natural value $\approx 1/\gamma$. In short, high brightness is impossible without a strict control of the electron beam geometry by the magnet lattice.

This requirement characterizes a synchrotron source with respect to other particle accelerators. The technology to control high-energy electrons was originally developed for research in the domain of elementary-particle physics. Synchrotron sources borrowed this technology and added to it sophisticated means to control the electron beam geometry.

At present, the transverse beam size is kept below a few thousandths of a millimeter over a ring circumference of hundreds of meters. The angular deviations are also extremely small. All these achievements are summarized by saying that the storage ring has a 'low emittance'.

As an electron circulates around the ring, it is forced to emit electromagnetic waves by the undulator, bending magnets and wigglers. The emitted waves carry energy, therefore their emission decreases the electron energy. Without some correcting action, the electron would rapidly lose energy and become unable to circulate in the ring.

This problem is eliminated by devices called 'radiofrequency (RF) cavities'. An electron passing through an RF cavity is 'kicked' by the cavity electric field to compensate the energy it lost by emitting waves. The 'kicking' cannot be indiscriminately applied to all passing electrons, otherwise those that did not lose energy would be accelerated beyond the energy required to circulate within the ring. As electrons enter an RF cavity, the cavity field must be 'off' during the passage of lossless electrons, and then turn 'on' for the arrival of the electrons that are retarded because of the emission of waves.

This synchronization implies (Fig. 1.7) that the electrons cannot circulate around the ring as a continuous flow but must form discrete bunches. The cavity applies a pulsed electric field for each passage of a bunch, to catch the arrival of its slowest electrons. Suppose that only one bunch circulates around the ring: the time period for a complete turn is given by the ring circumference—typically, 100 m or more—divided by the electron speed $\approx c = 3 \times 10^8$ m/s.

Thus, the period is of the order of $\approx 10^2/(3 \times 10^8) \approx 3 \times 10^{-7}$ s, corresponding to a frequency of $\approx 1/(3 \times 10^{-7}) \approx 3 \times 10^6$ s^{-1}. This is in the RF range, hence the need for an RF cavity. If several electron bunches circulate in the ring, then the 'kicking' frequency of the cavity increases but remains in the RF range.

The electron 'bunching' causes an important property of synchrotron sources: their emission is not continuous in time but pulsed. A beamline receives synchrotron light only when an electron bunch passes in front of it. The time structure of synchrotron light is thus shaped as shown schematically in Fig. 1.7: a series of pulses of duration δt, separated by 'dead times' of duration Δt.

The pulse duration is the length of an electron bunch divided by the electron speed $\approx c$; this gives δt-values down to the 10^{-9} s ('nanosecond') range or less ('picosecond' range). The dead time Δt corresponds to the distance between two adjacent bunches.

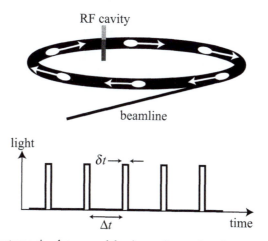

Fig. 1.7 Top: the electrons circulate around the ring as discrete bunches. When a bunch enters the RF cavity, a pulsed electric field is applied to 'kick' the electrons at the end of the bunch, i.e. the electrons that lost energy by emitting synchrotron light while turning around the ring. Bottom: synchrotron light is present in a beamline every time an electron bunch passes its front end. Thus, the light consists of a series of pulses of duration δt, separated by 'dead times' of duration Δt.

Some interesting synchrotron techniques exploit the pulsed time structure. For example, a synchrotron pulse can excite fluorescence from a sample, and the fluorescence decay can be studied during the dead time between pulses.

This simplified pictures gives a rough idea of the overall structure of a synchrotron facility. A real facility is of course more complicated: it must include, for example, equipped space to house staff and users, support laboratories to prepare experiments, construction shops for mechanical, vacuum and electronic components, a shielding system to prevent the leakage of potentially dangerous radiation, very sophisticated computer systems to control the operation of the ring and of the beamlines—and to handle the very large flow of experimental data produced by the facility.

1.2.1. A day in the life of a synchrotron

How does a real synchrotron facility operate? We can respond by going quickly through an entire cycle of operation—typically, one day. The starting point is immediately after the end of the previous cycle: this occurs when, many hours after the injection, the number of circulating electrons around the ring has decreased to an unacceptably low value.

There are several reasons for this decrease: some electrons are scattered out of their bunches by residual gas atoms—unavoidable even with ultrahigh vacuum. Others are lost because the RF cavity does not properly compensate their losses. Electron–electron interactions (e.g. the 'Touschek effect') can also eject electrons out of a bunch.

In a well-built storage ring, such losses are moderate: the injected electron beam has a long 'lifetime' (typically 20–40 hours or more). At a certain point, however, the synchrotron light emission (which is proportional to the number of circulating electrons) falls below a minimum acceptable level: a new operation cycle must be started.

The first step is the elimination of the old electron beam. Then, electrons are produced in the injector, pre-accelerated (in most facilities to the final required energy) and sent through an injection line into the main ring. The electrons' orbits are then optimized, and the new beam is ready for the users.

The beamlines—which typically were closed as a precautionary measure during the injection process—are opened again, and the experiments re-start. They can go on without interruptions until the end of the new operation cycle.

Most rings operate continuously, 24 hours per day and 365 days per year. This is needed to optimize the return on the construction investment, and also required by the demand that greatly exceeds the available capacity in almost all synchrotron facilities. However, some down time is required for maintenance, repairs and improvements of the ring and of the beamline instrumentation. The generally accepted minimum target for the beamtime actually delivered to the users is 5000 hours/year.

1.3. Controlling the source parameters? No problem!

We should now possess a general idea of a synchrotron facility and of the way it works. We can thus resume the discussion of the source properties. The next important topic is how to adjust the source parameters as required for each specific application. For example, many applications require a variable wavelength.

Fig. 1.8 A new day in the life of a synchrotron begins.

Considering eqn 1.4, changing the wavelength might appear difficult: the emitted λ_L of an undulator is determined by the (fixed) period L of the magnet array and by the γ-parameter, corresponding to the electron energy. In principle, one could modify the emitted wavelength by changing γ. But this is unrealistic: the storage ring supplies light to many beamlines simultaneously, and a change in electron energy would affect all of them.

To find a viable solution, we must analyze the details of the undulator emission mechanism. The γ-parameter in eqn 1.4 corresponds to the energy of the electron as it moves along a straight trajectory in the 'forward' direction. But the undulator causes small oscillations in the transverse direction.

The Lorentz force cannot change the energy of the electrons, therefore their speed remains constant. Thus, the creation of a velocity component in the transverse direction requires a decrease in the magnitude of the forward velocity component. In turn (eqn 1.4), this decreases the 'forward-motion' electron energy and the corresponding effective γ-value.

In conclusion, by changing the magnitude of the transverse oscillations we can change the effective γ-value that must be used in eqn 1.4 to evaluate the emitted wavelength. The magnitude of the transverse oscillations can be controlled by changing the strength B of the periodic magnetic field of the undulator—for example by increasing or decreasing the 'gap' between the magnet poles. In summary, by changing B the wavelength can be tuned to the desired value, or even continuously scanned during an experiment.

Quantitatively speaking (Inset C), the average transverse speed is proportional to B. As a consequence, the effective value of $1/\gamma^2$ in eqn 1.4 becomes $(1/\gamma^2)(1 + K^2/2)$, where K is a parameter proportional to B:

$$K \approx 0.934 \times B[\text{T}] \times L[\text{cm}] . \tag{1.7}$$

Therefore, the corrected version of eqn 1.4 is:

$$\lambda_L \approx \frac{L}{2\gamma^2}\left(1+\frac{K^2}{2}\right).$$

(1.8)

Equation 1.8 implies the possibility of changing the emitted wavelength λ_L by changing B and therefore K.

There exists another way to obtain different wavelengths from an undulator. Equation 1.8 is strictly valid for emission along the axis of the undulator (the x-axis). If we observe the emission at an angle θ slightly off-axis (Fig. 1.9), then λ_L changes with respect to the on-axis value.

Why? Basically, because the apparent source velocity that causes the Doppler shift of the wavelength changes with the angle θ. Equations 1.4 and 1.8 are valid for the on-axis Doppler shift, $\theta = 0$. In general, instead:

$$\lambda_L \approx \frac{L}{2\gamma^2\left(1+\frac{K^2}{2}+\gamma^2\theta^2\right)}.$$

(1.9)

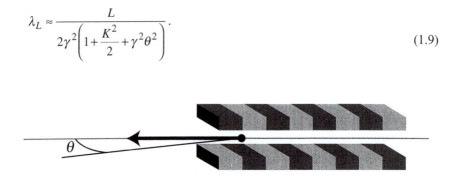

Fig. 1.9 Undulator emission at an off-axis angle θ in the undulation (horizontal) plane.

Inset C: Additional details about undulators

Transverse oscillations and wavelength tunability
We can easily estimate the effects of the transverse (undulation) velocity u_y on the effective γ-factor. Since the magnitude u of the velocity stays constant (the magnetic Lorentz force does not produce work), the forward component $u_x^2 = \sqrt{u^2 - u_y^2}$ is smaller than u.

The γ-factor definition (eqn A12) gives

$$\frac{1}{\gamma} = \sqrt{1 - \frac{u^2}{c^2}}, \text{ thus } \frac{1}{\gamma^2} = 1 - \frac{u^2}{c^2}.$$

The 'effective' γ-value must be calculated by replacing u with u_x, so that the above expression for $1/\gamma^2$ changes to:

$$1 - \frac{u^2 - u_y^2}{c^2} = \left(1 - \frac{u^2}{c^2}\right) + \frac{u_y^2}{c^2} = \frac{1}{\gamma^2} + \frac{u_y^2}{c^2} = \left(\frac{1}{\gamma^2}\right)\left(1 + \frac{\gamma^2 u_y^2}{c^2}\right).$$

Relativistic mechanics shows that the transverse (undulation) speed u_y is, on average, proportional to B/γ: $u_y = bB/\gamma$ (where b is a proportionality constant). Therefore, the expression for the effective $1/\gamma^2$ can be written as:

$$\frac{1}{\gamma^2}\left(1 + \frac{b^2 B^2}{c^2}\right).$$

This result is equivalent to eqn 8 if one identifies the K-parameter as

$$K = \sqrt{2}\,\frac{bB}{c}. \tag{C1}$$

This analysis qualitatively justifies eqns 1.7 and 1.8. A more detailed and less approximate theoretical treatment would give the exact expression of eqn 1.7 for the K-parameter.

Angular dependence of the wavelength

Why does the detected undulator wavelength change off-axis as a function of the angle θ? The response is that the Doppler shift is not the same in a generic θ-direction and along the x-axis ($\theta = 0$). In fact, the Doppler shift is caused by the source velocity, and the velocity component changes with the direction.

We can limit the analysis to very small values of θ since the undulator emission is highly collimated. In that limit, the high-speed Doppler-shift expression of eqn A15 valid for $\theta = 0$ can be generalized to $\theta \neq 0$ as follows:

$$\lambda_L = \frac{\lambda_e}{2\gamma\left(1 + \gamma^2\theta^2\right)}. \tag{C2}$$

Equation C2, combined with the undulation-effect equation (eqn 1.8), approximately gives eqn 1.9.

1.3.1. Higher harmonics, bandwidth, real collimation

Equation 1.9 defines the so-called 'fundamental' wavelength λ_L emitted by an undulator. In reality, an undulator emits other wavelengths besides λ_L.

First of all, like for the sounds produced by a musical instrument, the 'fundamental' emission can be accompanied by 'higher harmonics', whose wavelengths are integer fractions of the fundamental one:

higher harmonic wavelengths $= \lambda_L/n$, with $n = 1,2,3,4...$ (1.10)

(where $n = 1$ gives the fundamental wavelength). For small angular deviations from the forward direction ($\theta \to 0$) only odd harmonics ($n = 1, 3, 5 \dots$) are detected. Otherwise, even harmonics are also present.

Consider now the emission of one of the wavelengths of eqn 10. In reality, we cannot have just 'one' wavelength: each of the values of eqn 1.10 is really the center of a narrow wavelength *band*. In particular, the fundamental emission consists of a narrow band (Fig. 1.10) of width $\approx \Delta\lambda_L$, centered at λ_L.

The bandwidth $\Delta\lambda_L$ is determined by phenomena similar to those occurring for a diffraction grating in optics: the periodic magnet array of the undulator acts, in a sense, as a grating, which is a periodic array of 'lines'. We can thus borrow the relevant equation from optics: the 'relative' bandwidth (resolution function) for the nth harmonic is:

$$\frac{\Delta\lambda_L}{\lambda_L} = \frac{1}{nN},$$ (1.11)

where N is the total number of periods of our undulator grating.

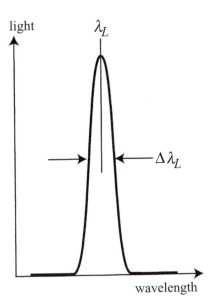

Fig. 1.10 Emitted wavelength band around the fundamental wavelength λ_L.

The final point about undulators is a nice surprise: the real angular spread of an undulator is even smaller than the already very small intrinsic value $\approx 1/\gamma$! In fact, angular deviations from the undulator axis correspond (eqn 1.9) to wavelength changes. But such changes cannot exceed the 'grating' bandwidth of eqn 1.11 without compromising the overall constructive interference that produces the undulator emission. The consequence (Inset D) is that the real angular spread for the fundamental wavelength is not $1/\gamma$ but:

$$\delta\theta \approx \frac{1}{\gamma\sqrt{N}}.$$
(1.12)

Since the number of periods N must be larger than one, this corresponds to an angular spread smaller than $1/\gamma$.

Inset D: Actual angular spread of the undulator emission

According to eqn C2, an angular deviation $\delta\theta$ would change the fundamental emitted wavelength with respect the on-axis value of eqn 1.4, $\lambda_L \approx L/(2\gamma^2)$. The new value is:

$$\lambda_L + \Delta\lambda_L \approx \frac{\lambda_e}{2\gamma}\left[1+\gamma^2(\delta\theta)2\right] \approx \frac{L}{2\gamma^2}\left[1+\gamma^2(\delta\theta)2\right].$$

The corresponding relative change $\Delta\lambda_L/\lambda_L$ is

$$\frac{\Delta\lambda_L}{\lambda_L} \approx \frac{(\lambda_L+\Delta\lambda_L)-\lambda_L}{\lambda_L} \approx \frac{\left(\dfrac{L}{2\gamma^2}\right)\left[1+\gamma^2(\delta\theta)^2\right]-\dfrac{L}{2\gamma^2}}{\dfrac{L}{2\gamma^2}} = \gamma^2(\delta\theta)^2.$$

On the other hand, this value of $\Delta\lambda_L/\lambda_L$ must be smaller than the 'natural' bandwidth given by eqn 1.11, whichfor the fundamental wavelength is $1/N$. Thus:

$$\gamma^2(\delta\theta)^2 \le 1/N,$$

corresponding to the maximum angular spread of eqn 1.12.

> **Undulators: essential summary**
> - The peak emission wavelength of an undulator is determined by the undulator period, shortened first by the Lorentz contraction and then by the Doppler shift.
> - The corresponding shortening factors must take into account the angular dependence of the Doppler effect and the effect of the magnetic-field-induced electron undulations.
> - The 'natural' bandwidth is given by the 'diffraction grating' effect of the series of magnets in the undulator.
> - The angular spread is determined by the fact that the corresponding wavelength spread cannot exceed the 'natural' bandwidth.

1.4. Bending magnets as synchrotron light sources

Having completed our basic discussion of undulators, we are now ready to consider another important type of synchrotron-light-emitting devices: the bending magnets. In the early days of synchrotron research, only bending magnet sources were used. The emission of light was a byproduct: the main function of the bending magnets was to keep the electrons in closed trajectories around the ring. Therefore, the source characteristics could not be easily optimized.

The technical progress of the subsequent years yielded better synchrotron sources than the bending magnets, notably the undulators. Bending magnet sources, however, still play an important role in synchrotron research. Even today, the majority of the synchrotron beamlines receive light from bending magnets. It is important, therefore, to be familiar with the basic properties of such sources.

What, for example, are the wavelengths emitted by a bending magnet? To answer, we must briefly analyze the emission mechanism. This is related to the fact that the electrons deflected by a bending magnet are accelerated electric charges—with centripetal acceleration caused by the Lorentz force* of the magnets.

Imagine (Fig. 1.11) an electron circulating around the ring seen from the point of view in the plane of the ring. When subjected to the action of a bending magnet, the electron moves along a portion of a circular trajectory. From the side, it looks like a charge oscillating along a straight line—as in a radio antenna. This oscillating charge emits electromagnetic waves whose frequency and wavelengths are determined by its oscillation frequency in the antenna.

(*) Incidentally, this notion may lead to a puzzle: when we use the *electron* reference frame rather than the laboratory frame, we might think that the electrons have zero acceleration, whereas acceleration is required for the emission of waves. This puzzle is solved by carefully considering the definition of the electron frame. This *is not* the frame moving as the electron: such a frame would be accelerated and therefore not an inertial frame as required by special relativity. The electron frame is instead the (inertial) frame that moves at a constant velocity equal to that of the electron when the light is emitted. In this frame, the electron has zero velocity but non-zero acceleration, and thus it emits electromagnetic waves.

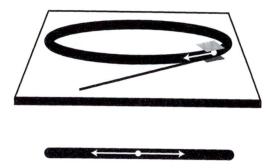

Fig. 1.11 Top: emission of synchrotron light by a bending magnet. Bottom: seen from the side, a circulating electron in the ring looks like an oscillating charge in a radio antenna.

In turn, the oscillation frequency corresponds to the angular speed of the electron along the circular trajectory. One can demonstrate (see Inset E) that the angular speed in the electron reference frame is

$$\omega_e = \frac{\gamma e B}{m_0},$$
(1.13)

corresponding to the oscillation frequency

$$\nu_e = \frac{\omega_e}{2\pi} = \frac{2\pi m_0 c}{\gamma e B},$$

and (see eqn 1.1) to the emitted wavelength:

$$\lambda_e = \frac{c}{\nu_e} = \frac{2\pi m_0 c}{\gamma e B}.$$
(1.14)

This is the wavelength seen in the *electron* reference frame. From the point of view of the laboratory, we must take into account the Doppler shift that decreases the observed wavelength by a factor $\approx 2\gamma$:

$$\lambda_L = \frac{\lambda_e}{2\gamma} = \frac{2\pi m_0 c}{2\gamma^2 e B}.$$
(1.15)

Note once again the appearance of the characteristic $2\gamma^2$ factor already found for undulators (eqn 1.4), whichreflects the combination of relativistic effects in the emission mechanism. The 'practical' version of eqn 1.14 is:

$$\lambda_L[\text{Å}] = \frac{5.3 \times 10^7}{B[\text{T}]} .$$ (1.16)

Consider, for example, a storage ring with 1 GeV energy ($\gamma \approx 2000$) and $B = 1$ T. The emitted wavelength would be $\lambda_L \approx 13$ Å.

1.4.1. Bending magnet emission spectrum

The wavelength λ_L of eqns 1.15 and 1.16 is not the *only* emitted wavelength, but the center of an emission band with a large bandwidth $\Delta\lambda_L$. The cause of this large bandwidth can be easily understood considering once again Fig. 1.10. Light can be detected only when an electron passes in front of the beamline. Thus, each electron produces a very short light pulse at each passage (note that this is *not* the observed pulse duration of Fig. 1.7, which is caused by all the electrons in a bunch and not by a single electron).

The short duration of the single-electron light pulse affects the wavelength bandwidth. We can understand why by analyzing (Fig. 1.11) a similar effect for sound waves.

A pure sound corresponds to a perfect oscillation with only one wavelength (and one frequency). On the other hand, the 'Fourier theorem' shows that any sound, no matter how complex, is always the superposition of pure sounds, i.e. of individual, well-defined wavelengths (and frequencies). This is why, for example, we can use a computer to synthesize complex sounds by combining pure sounds.

A similar Fourier-theorem property exists for other types of waves. For example, a beam of 'white' visible light is a superposition of individual pure colors, corresponding to specific and well-defined wavelengths.

There exists a general property of all Fourier combinations of wavelengths that we can easily visualize in the case of sounds (Fig. 1.12). A short pulse contains a broader band of frequencies (and wavelengths) than a long pulse. Quantitatively speaking, a pulse of duration Δt corresponds to a frequency bandwidth $\Delta\nu$ such that

$$\Delta t \Delta \nu \approx 1/(2\pi) .$$ (1.17)

Therefore, a short pulse Δt corresponds to a large frequency bandwidth $\Delta\nu$. Since frequency and wavelength are related to each other by eqn 1.1, a large frequency bandwidth also corresponds to a large wavelength bandwidth $\Delta\lambda$.

Equation 1.17 can be used to calculate the bandwidth $\Delta\lambda_L$ for a bending magnet. This is a rather simple exercise (Inset E), and the interesting result is $\Delta\lambda_L \approx \lambda_L$, so that

$$\frac{\Delta\lambda_L}{\lambda_L} \approx 1 .$$ (1.18)

We can roughly imagine the emitted spectrum of a bending magnet as a broad bell-shaped peak centered at λ_L and with approximate width $\Delta\lambda_L \approx \lambda_L$. A more sophisticated treatment shows that this simplified picture is not too far from reality—although the real spectrum is more complicated and a bit asymmetric rather than bell-shaped.

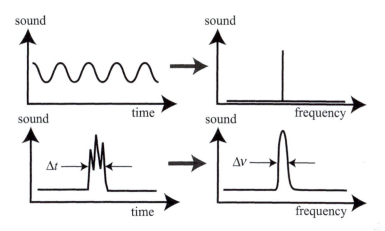

Fig. 1.12 Link between time structure and frequency bandwidth (Fourier theorem) for sound waves. Top: a pure sound corresponds to only one frequency. Bottom: a complex sound pulse of duration Δt is a superposition of different frequencies over a band of width $\Delta \nu$.

A similar conclusion is valid (Fig. 1.12) for the emission spectrum as a function of the photon energy, $E = h\nu = hc/\lambda$, rather than of the wavelength. A bell-shaped spectrum as a function of the wavelength corresponds in fact to a bell-shaped spectrum (Fig. 1.13a) as a function of $h\nu$. From the definition of the central wavelength λ_L (eqn 1.15), we can derive the central photon energy of this spectrum:

$$h\nu_L = \frac{hc}{\lambda_L} = \frac{\gamma 2eBh}{\pi m_0} .$$
(1.19)

The photon-energy bandwidth of the bell-shaped spectrum (Inset E) is:

$$\Delta h\nu_L = \frac{\gamma 2eBh}{\pi m_0} = h\nu_L .$$
(1.20)

The reader already familiar with synchrotron light might not recognize in the bell-shaped peak plot of Fig. 1.13a the familiar spectrum of a bending magnet source. This, in fact, is not the most frequently used plot: one normally shows this curve in a log–log scale rather than in a linear–linear scale. The corresponding shape of Fig. 1.13b is quite similar to the well-known bending-magnet spectrum, derived with a much more complicated and advanced electrodynamic treatment.

Note that the bending magnet emission extends over many photon energy (and wavelength) domains besides ultraviolet light and X-rays. These include, for example, infrared and visible light. From this broad emission, specific wavelengths can be filtered with *ad hoc* devices, as required for individual applications.

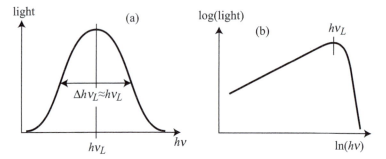

Fig. 1.13 (a) Approximate shape of the emitted light from a bending magnet as a function of the photon energy; (b) the corresponding log–log plot closely resembles the well-known synchrotron spectrum predicted by an advanced theoretical treatment.

Quite often, different parameters are used instead of hv_L and λ_L to characterize the bending magnet emission, namely the 'critical photon energy' hv_c and the corresponding 'critical wavelength' λ_c. These are defined by requiring that equal amounts of synchrotron light are emitted above and below hv_c. Theory shows that

$$hv_c = \frac{3}{4}hv_L = \frac{3\gamma^2 eBh}{4\pi m_0}; \qquad \lambda_c = \frac{4}{3}\lambda_L = \frac{4\pi m_0 c}{3\gamma^2 eB}. \tag{1.21}$$

The corresponding practical definitions are

$$hv_c \approx 1.7 \times 10^{-4}\gamma^2 B[\text{T}]; \qquad \lambda_c[\text{Å}] \approx \frac{7 \times 10^7}{\gamma^2 B[\text{T}]}. \tag{1.22}$$

For example, a ring with 2 GeV energy ($\gamma \approx 4000$) and 4-T magnets would have a critical photon energy $hv_c \approx 2.7 \times 10^3$ eV and a critical wavelength $\lambda_c \approx 4.4$ Å.

Inset E: Bending magnets and their emission

Central emitted wavelength

The 'central' emitted wavelength of a bending magnet is related to the angular speed of the electron motion caused by the magnet. This angular speed can be easily derived: in the laboratory frame, the electron acceleration is the centripetal acceleration, equal to the product of the angular speed times the tangential speed, $\omega_L u$. Multiplied by the (relativistic) electron mass, $m = \gamma m_0$, this must be equal to the Lorentz force, euB:

$$\gamma m_0 \omega_L u = euB,$$

and therefore:

$$\omega_L = \frac{eB}{\gamma m_0}.$$ (E1)

A similar result is valid for the reference frame of the electron, but according to relativity the Lorentz force is replaced by an electrostatic force of magnitude γeuB and the mass is the electron rest mass m_0; thus

$$\omega_e = \frac{\gamma eB}{m_o},$$ (E2)

confirming eqn 1.13, whichwas used to derive the central emitted wavelength (eqns 1.15 and 1.16).

Wavelength bandwidth

According to eqn 1.17, $\Delta t\,\Delta v \approx 1/(2\pi)$, the emission frequency bandwidth Δv must be evaluated by calculating the duration Δt of a light pulse produced by an electron passing through the bending magnet. This is an interesting exercise that enables us to understand better the underlying relativistic mechanisms.

Imagine (Fig. E-1) a circulating electron like a miniature 'torchlight', whose emitted light has the angular spread $\approx 1/\gamma$ typical of synchrotron emission. The detection of a light pulse begins when the front edge of this light cone reaches the detector. Assume that the corresponding distance between the emission point and the detector is A. If we call $t = 0$ s the emission time, then the pulse at the detector begins at the

$$\text{detection start time} = \frac{A}{c}.$$

The same pulse ends when the cone of light has rotated by an angle equal to its width, $1/\gamma$. Since (eqn E1) the angular speed in the laboratory reference frame is $eB/\gamma m_0$, the time required for this rotation is $t = (1/\gamma)/(eB/\gamma m_0) = m_0/eB$.

This *is not*, however, the pulse duration: during the time period t the electron (source) moves ahead by $\approx ut = um_0/eB$, decreasing its distance from the detector from the value A to $\approx A - um_0/eB$.

Therefore, the end-cone emission occurs at the time $t = m_0/eB$, but its detection occurs $(A - um_0/eB)/c$ seconds later, i.e. at the

$$\text{detection end time} = \frac{m_0}{eB} + \frac{A - \dfrac{um_0}{eB}}{c} = \left(\frac{m_0}{eB}\right)\left(1 - \frac{u}{c}\right) + \frac{A}{c}.$$

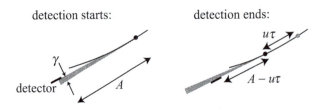

detection starts:

detection ends:

Fig. E-1 Duration of a detected pulse emitted by an electron passing through a bending magnet.

The pulse duration is of course the difference between the detection end time and the detection start time:

$$\Delta t = \left(\frac{m_0}{eB}\right)\left(1 - \frac{u}{c}\right). \tag{E3}$$

On the other hand, for $u \approx c$ we can approximate* $(1 - u/c)$ as $1/(2\gamma^2)$ and:

$$\Delta t = \frac{m_0}{2\gamma^2 eB}.$$

The corresponding frequency bandwidth (according to the Fourier theorem, eqn 1.17: $\Delta v \Delta t \approx 1/(2\pi)$) is

$$\Delta v = \frac{1}{2\pi\Delta t} \approx \frac{2\gamma^2 eB}{2\pi m_0}, \tag{E4}$$

which, considering eqn 1.15, simply corresponds to

$$\Delta v = \frac{c}{\lambda_L}. \tag{E5}$$

From the frequency bandwidth we can extract the wavelength bandwidth using the wavelength–frequency relation of eqn 1.1, $v = c/\lambda$. Taking the differential of this expression, we obtain for the bandwidth magnitude $\Delta v \approx c\Delta\lambda/\lambda^2$; therefore, the bandwidth $\Delta\lambda_L$ around the 'central' wavelength λ_L is:

$$(*) \ 1 - \frac{u}{c} = \frac{\left(1 - \frac{u}{c}\right)\left(1 + \frac{u}{c}\right)}{1 + \frac{u}{c}} = \frac{1 - \frac{u^2}{c^2}}{1 + \frac{u}{c}} = \frac{1}{\gamma^2\left(1 + \frac{u}{c}\right)} \approx \frac{1}{2\gamma^2}.$$

$$\Delta\lambda_L \approx \frac{\lambda_L{}^2 \Delta\nu}{c} = \lambda_L .$$

This result corresponds to eqn 1.18.

1.4.2. Bending magnet flux and brightness: a truly powerful source

Is a bending magnet bright? Indeed it is, but for quite different reasons than an undulator. Consider the angular spreads $\delta\theta_y$ and $\delta\theta_z$ that determine the brightness (eqn 1.3). For a bending magnet, one must distinguish between the horizontal and vertical directions.

Vertically, the spread is still determined by the intrinsic $1/\gamma$ factor and by the electron beam geometry. Horizontally, there is no built-in limitation to the angular spread: Fig. 1.10 shows that the emission direction changes as the electron moves. On the other hand, each beamline only accepts a limited horizontal angle, which thus defines the practical horizontal spread.

As we have seen, a key difference between undulators and bending magnets is that the undulator emission occurs in a narrow wavelength bandwidth, whereas the bending magnet emits over a broad band. Therefore, an undulator is an ideal source when a specific wavelength is required. A bending magnet (or a wiggler) is a better source when 'white' synchrotron light is required, i.e. light with many different wavelengths.

But we can also obtain narrow-band 'monochromatic' light from a bending magnet beamline. This is done by inserting along the beamline a filtering device called a 'monochromator' to reject all other wavelengths. The filtered-out light is not used, thus bending magnets are less efficient than undulators as monochromatic sources. On the other hand, monochromatized bending magnets can yield light over a wide wavelength range, whereas the undulator wavelength-tunability, based on eqn 1.9, is limited.

We should note that many undulator beamlines also include monochromators. In fact, the undulator bandwidth (eqn 1.10) is not small enough for most applications and requires additional filtering.

The final comment about the emission of different types of synchrotron sources is that the emitted flux of light is amazingly high for all of them—undulators, bending magnets and wigglers. For example, the heat equivalent of an undulator challenges what one finds at the surface of the sun. What causes all this power?

We have seen that electromagnetic waves are emitted by accelerated electric charges. Theory shows that the emitted power is proportional to the square of the acceleration. For an electron in a curved trajectory, the acceleration is proportional to the angular speed, which in turn (eqn 1.13) is proportional to γB. Therefore:

$$\text{emitted power} = \text{constant} \times \gamma^2 B^2 . \tag{1.23}$$

For constant B, the emission is thus proportional to the square of the γ-factor—that is, to the square of the relativistic electron energy. Since γ is large, each electron emits a lot of light. Furthermore, the total output of a synchrotron source is the sum of the emissions of all the circulating electrons, and therefore proportional to the electron beam current.

These combined factors can produce an extremely high power. For example, the total bending-magnet emitted power of a 1 GeV synchrotron, with $B = 1.7$ T and a circulating electron current of 100 mA, is $\approx$4.5 kW; a 2.5 GeV ring with $B = 1.2$ T and the same current would give $\approx$50 kW.

Bending magnets: essential summary

- The small vertical angular spread of bending magnet emission is again determined by the electron beam geometry and by the natural angular spread $1/\gamma$, which in turn is due to the fact that the emitted waves are 'projected forward' by the motion of the source (the electron). This forward-projection impact on angular collimation is extreme because of relativistic effects.
- The peak emission wavelength of a bending magnet is determined by the angular speed of the electron motion. In turn, the angular speed is determined by the bending magnet field strength and is affected by relativistic phenomena.
- The broad wavelength bandwidth of a bending magnet reflects the short duration in time of the light pulse produced when an electron passes in front of a beamline.

1.5. Wigglers

A wiggler is a device similar to an undulator: a periodic series of magnets inserted along a straight section of the ring. The difference is that the B-field strength is higher, and so is the K-parameter according to eqn 1.7.

A higher B-value (and K-value) causes larger transverse undulations of the electrons, with important consequences (Fig. 1.14). The fundamental wavelength given by eqn 1.9 increases with K. At the same time, the emitted light intensity is shifted from the fundamental wavelength to higher harmonics (Equ. 1.10) at shorter wavelengths. Overall, most of the light is thus emitted at *shorter* wavelengths as the B-field increases.

What causes this intensity shift to higher harmonics? Consider Fig. 1.14. For small undulations (undulator regime) the light cone emitted by an electron is continuously detected during the entire passage of the electron through the magnet array. Thus, the detected light as a function of time is a rather long pulse.

This pulse is the *coherent* result of all electron undulations, producing the fundamental undulator wavelength of eqn 4, $\lambda_L = L/2\gamma^2$. The bandwidth $\Delta\lambda_L$ around the fundamental wavelength is caused by the undulator 'grating' effect, and is quite narrow.

As the B-field strength increases, the source moves from the 'undulator' regime to the 'wiggler' regime (Fig. 1.14b–1.14d). The undulations become larger and the angular deviations make it impossible for the emitted light cone to be continuously detected. The result is a series of short light pulses rather than a single long pulse.

According to the Fourier theorem (eqn 1.17, $\Delta t\, \Delta v \approx 1/(2\pi)$), the bandwidth around each emitted wavelength increases. For very high B-values (and K-values), the individual peaks merge into a broad-band continuum. This tendency is enhanced if the number of periods in the magnet array is small.

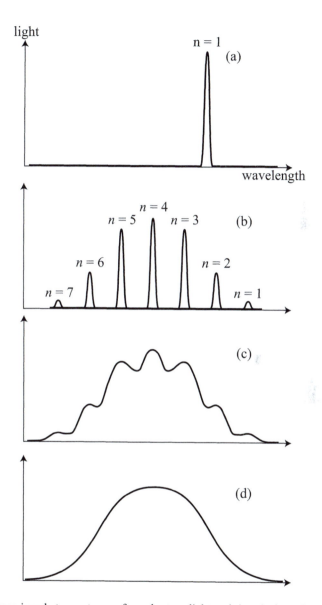

Fig. 1.14 Comparison between types of synchrotron-light-emitting devices. (a) An undulator with a very weak B-field (and K-parameter) emits a narrow band around the fundamental wavelength ($n = 1$). (b) As B and K increase, the fundamental wavelength ($n = 1$) shifts to higher values, but the weight of higher harmonics increases too. Therefore, the overall emission intensity shifts to *smaller* wavelengths. (c) For larger and larger B- and K-values (and/or for a small number of periods) the individual bands are broadened. (d) As a limit case (pure wiggler), there is only one broad band. This is also the emission of a bending magnet, and a wiggler is in fact equivalent to a series of bending magnets.

In the limit case of very large B-values (Fig. 1.14d), the individual undulations produce light as sources independent of each other rather than coherent, and the wiggler is practically equivalent to a series of bending magnets. The central emitted wavelength is not determined by the magnet array period (eqn 1.4) but by the angular speed (eqn 1.15), which gives shorter wavelength values.

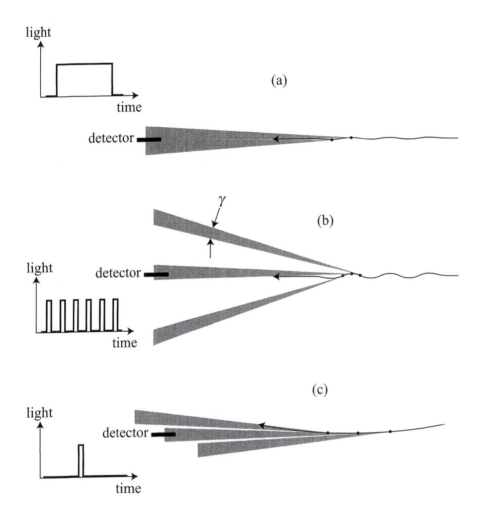

Fig. 1.15 Time structure of the light emitted by an individual electron passing through different types of synchrotron light sources. (a) An undulator with a weak B-field (and K-parameter) causes small angular deflections of the electron trajectory with respect to a straight line. As a consequence, the light cone (of angular spread $\approx 1/\gamma$) can continuously illuminate the detector, producing a long pulse of light. (b) As the B-field increases and the source moves to the wiggler regime, the larger undulations produce a series of short pulses. (c) For a bending magnet there is a single short pulse for each passage of the electron.

This explains why in the transition from the undulator mode to the wiggler mode the emission progressively shifts to higher harmonics. This reflects the increasing prevalence of the central bending-magnet wavelength over the fundamental undulator wavelength.

From Fig. 1.15 we can understand what marks the transition between the undulator regime and the wiggler regime. The undulator-regime condition of continuous illumination (and coherent light emission) is satisfied if the undulation-caused angular deviations are smaller than the light-cone width, $\approx 1/\gamma$. One can demonstrate that this means $K < 1$, since the K-parameter is the ratio between the undulation-caused angular deviation and $1/\gamma$.

Roughly speaking, $K < 1$ corresponds to an undulator with a strong and narrow fundamental-wavelength emission. K much larger than 1 corresponds to a wiggler, equivalent to a series of bending magnets. There is a broad transition regime (Figs. 1.14b, 1.14c) for which the emission still consists of peaks rather than of a broad band.

1.6. Polarization

Polarization is one of the most important characteristics of synchrotron light. Its practical applications are numerous and particularly relevant for chemistry and the life sciences. We should then briefly explain what it is and why it is present for synchrotron light.

An electromagnetic wave is, as we have mentioned several times, a propagating perturbation of the electromagnetic field. The electromagnetic field includes an electric field vector and a magnetic field vector. The word 'vector' means that these fields are characterized both by their strength and by their direction.

In an electromagnetic wave, the propagating perturbations of the electric and magnetic fields occur in directions perpendicular to the wave motion. Thus, electromagnetic waves are 'transverse' waves. Furthermore, the electric field perturbation and the magnetic field perturbation are perpendicular to each other—see Fig. 1.16.

These properties do not completely identify the direction of the electric field (or magnetic field), which can be any direction transverse to the wave motion. In certain cases, however, this direction is fixed—see Fig. 1.16b. One calls the corresponding waves 'linearly polarized'. In other cases, at a fixed site the electric field of the wave rotates as a function of time, with the tip of its representative arrow following a circle or an ellipse, and the corresponding wave is called 'circularly' or 'elliptically' polarized.

Imagine now an electron circulating in a synchrotron source and subject to the action of a bending magnet. We have seen (Fig. 1.11) that observed from the side (ring plane) the electron looks like an oscillating charge in an antenna. Not surprisingly, the bending-magnet emission is linearly polarized when detected in the plane of the ring.

Let us now adopt a different point of view slightly above or below the ring plane. We see the circulating electron moving along an elliptical curve; thus, we can understand why the out-of-plane emission is elliptically polarized.

Many life-sciences experiments require intense non-linear polarization, and the out-of-plane bending-magnet emission is not sufficient. The problem can be solved by special undulators and wigglers such as the 'helical devices'. The emission of a standard undulator or wiggler is linearly polarized when seen from a point of view in the undulation plane.

Helical devices produce an electron motion more complicated than a simple 'undulation' in a plane. For example, 'elliptical wigglers' force the electrons into a spiral trajectory with an elliptical transverse section, thus producing controllable non-linear polarization.

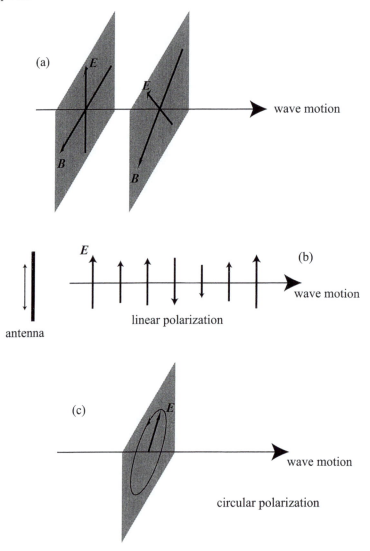

Fig. 1.16 Polarization. (a) The magnetic (**B**) and electric (**E**) field perturbations of an electromagnetic wave occur in directions perpendicular to each other and also perpendicular to the wave motion; this, however, is not sufficient to fix the two directions. (b) If, on the other hand, the electric field and magnetic field directions are fixed, then the wave is linearly polarized; this is the case of an antenna-emitted wave antenna in a direction perpendicular to the antenna. (c) If the electric-field direction at a given site rotates as a function of time, then the wave is circulary or elliptically polarized.

1.7. Coherence

Coherence is a special and important property of certain light sources. Coherent sources of visible light—such as lasers—have been available for a long time. Coherent X-rays, on the contrary, were not available until the advent of advanced synchrotron facilities. The impact of their recent arrival is extremely important for many different domains.

The notion of coherence can be understood with simple arguments. Visible light sometimes produces phenomena known as 'diffraction' or 'interference'—for example, the rainbow interference patterns of thin water layers after a rainfall. In most cases, however, such phenomena are not observed. For example, a standard lamp in our home cannot, except in special cases, produce them.

A 'coherent' wave is, loosely speaking, a wave that can produce observable interference and diffraction phenomena. To quantify this notion, consider for example the interference phenomenon produced by two parallel slits—Fig. 17a. In the interference pattern, an intensity maximum is detected whenever the path difference δx behind the slits between the two rays is an integer multiple of the wavelength:

$$\frac{\delta x}{\lambda} = 1, 2, 3, \ldots ; \tag{1.24a}$$

on the other hand, whenever:

$$\frac{\delta x}{\lambda} = \frac{1}{2}, \frac{3}{2}, \frac{5}{2}, \ldots , \tag{1.24b}$$

there is destructive interference and a dark fringe in the interference pattern.

These properties emphasize the importance of the 'phase factor' x/λ and of the difference between the phase factors of the two rays, $\delta x/\lambda$. Specifically, constructive interference and destructive interference correspond to integer and semi-integer values of the phase factor difference.

The result is a series of easily observable bright and dark fringes. This conclusion, however, is valid if the phase factor x/λ is well defined when the waves reach the two slits. In the case of Fig. 1.17a, this is certainly true if the source is a point source that emits only one wavelength.

The situation changes if the source is *of finite size* rather than being a point, and/or if it emits a *finite wavelength bandwidth* $\Delta\lambda$ rather than a single wavelength. Then, the fringes may no longer be visible—which means a loss of source coherence.

1.7.1. Transverse coherence

Consider first the effects of a finite source size (Fig. 1.17b). Each one of the two slits is illuminated by rays traveling along different paths and arriving at the slit with different phase factors x/λ. The maximum phase-factor difference occurs between rays from the two ends of the source, which are at a distance σ_y from each other.

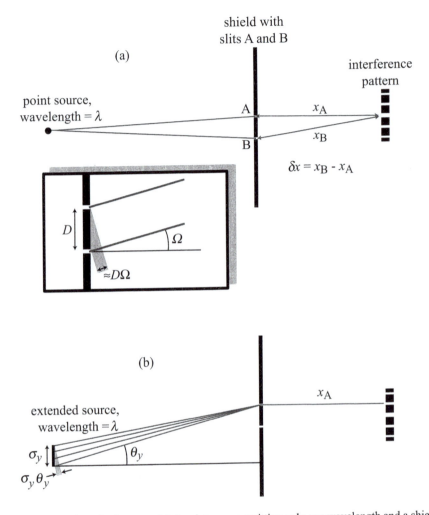

Fig. 1.17 The notion of coherence: (a) A point source emitting only one wavelength and a shield with two slits produce an interference pattern. The two rays passing through the two slits A and B arrive at a given point of the pattern after different paths x_A and x_B. The conditions for constructive or destructive interference are determined by the path difference δx and by the corresponding phase-factor difference $\delta x/\lambda$, according to eqns 1.24a and 1.24b. The inset shows the link between δx, the slit distance D and the angle Ω. Equations 1.24a and 1.24b, however, assume that there is no phase-factor difference between the two rays when they reach the slits. This is true for a point source symmetrically placed with respect to the two slits and emitting only one wavelength, as shown in the figure. A laterally displaced point source would produce a fixed phase-factor difference between the two rays and displace the interference pattern without destroying it. (b) On the other hand, an extended source of size σ_y at an illumination angle θ_y produces many rays that arrive at each of the two slits with different phase factors. The corresponding 'blurring' of the phase factor is $\sigma_y\theta_y/\lambda$. Similarly, a source whose emission occurs over a finite wavelength bandwidth also produces a phase-factor blurring. The wave is said to be *coherent* if the 'blurring' is small and the interference pattern is still visible.

If the illumination angle is θ_y, then the maximum path difference is $\Delta x \approx \sigma_y \theta_y$ and the corresponding phase-factor difference is $\sigma_y \theta_y / \lambda$. Thus, the different rays from different points of the source arrive at one slit with a 'blurred' distribution of phase factors, the 'blurring' being $\approx \sigma_y \theta_y / \lambda$.

Such a blurring affects the conditions for constructive and destructive interference and can wash out the interference pattern. With moderate blurring, the pattern is still visible—but what does 'moderate' mean? The answer is provided by eqns 1.24a and 1.24b: from constructive interference to destructive interference, the phase factor changes by 1/2. Thus, the phase-factor blurring $\sigma_y \theta_y / \lambda$ is 'moderate' if it is much smaller than 1/2, allowing a clear distinction between constructive and destructive interference:

$$\frac{\sigma_y \theta_y}{\lambda} \ll \frac{1}{2}. \tag{1.25}$$

This is the first condition for coherence, and is known as 'spatial coherence', 'lateral coherence' or 'transverse coherence'.

Equation 1.25 implies that only rays emitted within a limited angular range can produce diffraction or interference effects that require transverse coherence. The limit angles $\pm\theta_y$ derived from eqn 1.25 are $\pm\lambda/(2\sigma_y)$, corresponding to an angular interval $2\lambda/(2\sigma_y) = \lambda/\sigma_y$. Suppose that the overall angular spread of the source emission is $\delta\theta_y$: only a fraction $(\lambda/\sigma_y)/(\delta\theta_y) = \lambda/(\sigma_y\delta\theta_y)$ can produce phenomena that require coherence.

A similar condition would be valid for the perpendicular z-direction, further reducing this fraction by a factor $\lambda/(\sigma_z\delta\theta_z)$. Overall, only a fraction

$$C = \frac{\lambda^2}{\sigma_y \sigma_z \delta\theta_y \delta\theta_z} \tag{1.26}$$

of the source emission can contribute to interference (or diffraction) phenomena that require transverse coherence. This quantity C is called the 'transverse fraction of coherent power' of the source.

Note that the denominator in eqn 1.26 is exactly the same as that in eqn 1.3, which defined the source brightness. Thus, if the *brightness* is improved by decreasing the source size and angular divergence, then the *coherent power* is also increased.

The efforts in the 1980s and 1990s to enhance the brightness of synchrotron sources produced as a byproduct, for the first time in history, X-rays with high transverse coherence. For example, when Wilfried Hirt, Bruno Reihl, Lenny Rivkin, Werner Joho and myself conceived in 1990 the first notion of a Swiss synchrotron with very low emittance (what eventually became the Swiss Light Source or SLS), our main goal was brightness. The final result, however, was also a source with laser-like coherence.

What is the maximum possible level of coherent power? We can answer by noting that laterally-coherent X-rays could be extracted even from a large-size, divergent source. In fact, we could filter its emission with a shield having a narrow pinhole that would become the real source (Fig. 1.18); the source size σ_y would equal the pinhole size.

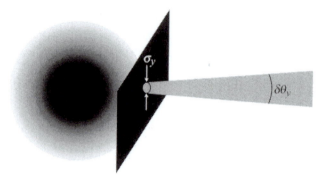

Fig. 1.18 A large-size source with large angular spread is converted into a coherent source by a shield with a pinhole. If the pinhole is very small, then diffraction effects determine the minimum angular spread of the waves originating from it.

This, however, is a very inefficient way to obtain high transverse coherence, since it wastes all the waves except for the tiny fraction passing through the pinhole. On the other hand, this pinhole-based approach makes it possible for us to understand the maximum reachable limit for transverse coherence.

When the pinhole width σ_y becomes too small, it causes a diffraction phenomenon that spreads the waves over an angular range $\delta\theta_y \approx \lambda/\sigma_y$. Thus, the product $\sigma_y\delta\theta_y$ cannot be smaller than the wavelength λ.

This is the fundamental 'diffraction limit' for the transverse coherence of any source. When a source reaches the diffraction limit for both the y- and z-directions, $\sigma_y\delta\theta_y = \lambda$ and $\sigma_z\delta\theta_z = \lambda$, according to eqn 1.26 it has coherent power $C = 1$, and thus is 'fully coherent'.

Note (eqn 1.26) that for a fixed source geometry the coherent power becomes smaller when the wavelength decreases. Thus, transverse coherence is *more difficult to achieve for short wavelengths*.

The most advanced synchrotron sources reach full (diffraction-limit) transverse coherence in a large portion of the emitted spectrum. For example, Elettra (Trieste) is fully coherent for λ down to ≈ 1000 Å, and the Swiss Light Source down to ≈ 100 Å. Not even lasers could be more coherent: synchrotrons are laterally coherent (at high λs) like lasers, although they are based on a different emission mechanism.

1.7.2. Longitudinal coherence

We must now consider the impact on coherence of a non-zero wavelength bandwidth $\Delta\lambda$. This 'blurring' of the wavelength causes a 'blurring' of the phase factor x/λ for the waves arriving at the two slits of Fig. 1.17a. The magnitude of the phase-factor blurring can be estimated by taking the derivative of the phase factor, $\partial(x/\lambda)/\partial\lambda = -x/\lambda^2$, and multiplying its magnitude by $\Delta\lambda$:

$$\text{phase - factor blurring} \approx \frac{x\Delta\lambda}{\lambda^2}.$$

Therefore, a wave that initially has no blurring becomes progressively more phase-factor blurred as it travels along a path of increasing length x. Once again, the observation of diffraction and interference effects requires the phase-factor blurring to be smaller than 1/2. This corresponds to a path $x < L_c$, where L_c is such that $L_c \Delta\lambda/\lambda^2 = 1/2$; therefore

$$L_c = \frac{\lambda^2}{2\Delta\lambda}.$$
(1.27)

Equation 1.27 defines the 'coherence length' L_c of the wave. In turn, the coherence length characterizes the wave 'longitudinal coherence' or 'temporal coherence'. Note that if $\Delta\lambda$ is infinitely small, then L_c becomes infinite.

In many applications, the practical longitudinal-coherence conditions are quite forgiving. Consider for example the two-slit interference of Fig. 1.17a with a point source of wavelength λ: the first dark fringe occurs for $\delta x = \lambda/2$ and the first-order bright fringe for $\delta x = \lambda$.

If the wavelength λ is 'blurred' and becomes a band of width $\Delta\lambda$, then the dark fringe is blurred by $\Delta(\delta x) = \Delta\lambda/2$. But the fringe is still visible if this blurring is reasonably smaller than the distance between the dark fringe and the bright fringe: $\Delta\lambda/2 < \lambda - \lambda/2 = \lambda/2$, which gives

$$\frac{\Delta\lambda}{\lambda} < 1.$$
(1.28)

This is a rather mild requirement. For example, eqn 1.18, is (almost) automatically satisfied by bending-magnet sources. And it is certainly satisfied by the much narrower bandwidth of undulators.

1.7.3. Why coherence?

What makes coherence so important, particularly in X-ray sources? To answer, we can propose a simplistic example that anticipates some of the applications of synchrotron light discussed later. Consider again the two-slit interference of Fig. 1.17a. The first dark fringe is observed for $\delta x = \lambda/2$. Defining D as the distance between the two slits, the inset in Fig. 1.17a shows that $\delta x \approx D\Omega$ and thus the first dark fringe corresponds to $\Omega = \lambda/(2D)$.

Therefore, by measuring λ and the angle Ω, we can derive the distance D. Similar considerations provide the background of all diffraction-based measurements of distances and positions in space. If, however, λ is large with respect to D, then the angle Ω increases too much and the measurement is no longer feasible. Roughly speaking, the value of λ must not be too far from the distance D that must be measured.

X-ray wavelengths range from 0.1 Å to 100 Å. The corresponding range of measurable distances includes the chemical bond lengths between atoms in solids and molecules.

Thus, X-rays can be used to measure distances and positions on the atomic scale: this is the background of fundamentally important techniques such as X-ray crystallography and X-ray scattering.

Coherence makes these techniques more powerful, and this explains why it is so important. Furthermore, we will see later in the book that X-ray coherence opens the door to a wide array of new applications, most notably in X-ray imaging and specifically in non-conventional radiology—as well as in X-ray holography.

2. The facility

The storage ring and its many sources of synchrotron light—undulators, wigglers and bending magnets—constitutes the core of a synchrotron facility. Such sources, however, would be useless without additional equipment. Quite often, for example, optical devices are needed to modify the spectral and geometrical characteristics of the emitted light beam and meet the requirements of specific applications. The back-end of each beamline must be equipped with an experimental chamber that includes different instruments, and interfaced with computers for controlling the experimental parameters and taking, processing, storing and exporting the data.

The many different applications of synchrotron light require a wide variety of instruments. A complete and detailed description of such instruments is well beyond our scope, and so we will limit the discussion to general issues and a to some examples of specific relevance to the audience of this book.

2.1. Beamlines

A synchrotron source (bending magnet, wiggler or undulator) is linked to the experimental chamber by a beamline that includes several optical components. Such devices are required to optimize the spectral characteristics of the synchrotron light beam as well as its geometry.

The optical devices for a synchrotron beamline are quite different from those used for visible light. For example, standard lenses do not work for X-rays and must be replaced by other types of focusing elements. A standard lens, in fact, is based on the refraction of light at glass–air interfaces. However, 'soft' X-rays are absorbed by all solid materials and in particular by glass. 'Hard' X-rays with shorter wavelengths can travel through glass, but their refractive index is very close to unity, and the corresponding refraction unsuitable for lenses.

As a consequence, with a few exceptions optical devices for X-rays are based on reflection rather than refraction or other mechanisms. The typical focusing devices are mirrors and diffraction gratings work in reflection rather than in transmission.

The reflection of X-rays is not immune from serious problems. The reflected intensity of X-rays sharply decreases when θ_g—the 'grazing' angle with respect to the surface—increases (see Fig. 2.1). This tendency is enhanced at short wavelengths—see Fig. 2.2. For hard X-rays, only beams with $\theta_g \approx 0$ (the grazing incidence) are reflected.

For ultraviolet light this phenomenon is not as marked as for X-rays: reasonably strong reflection occurs even near normal incidence ($\theta_g \approx \pi/2$). Roughly speaking, reflecting optical devices designed to work near normal incidence can be used if the wavelength does not exceed 40–50 Å—whereas a grazing-incidence geometry is required at shorter wavelengths. Normal-incidence reflection can be enhanced by multilayer coatings, as discussed later in the section on Schwartzschild lenses.

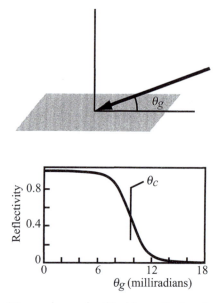

Fig. 2.1 Top: definition of the grazing angle of incidence, θ_g, for a beam of X-rays reaching a surface. Bottom: reflectivity of a gold surface as a function of the grazing angle of incidence for X-rays of wavelength $\lambda = 1.5$ Å. Note that the reflectivity decreases very rapidly as θ_g increases beyond the 'critical' value θ_c.

One of the most important functions of optical devices along a beamline is to focus and re-focus the X-ray beam. How can we, practically speaking, focus an X-ray beam? The simplest approach is to use a curved-surface mirror. We know that a parabolic mirror focuses a collimated beam of visible light into a point. The same effect is approximately produced by a spherical mirror. Mirrors of this kind can also be used at near-normal incidence to focus ultraviolet light.

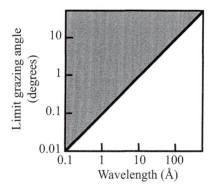

Fig. 2.2 A practical rule to evaluate the limit value of the grazing angle θ_g beyond which the reflection is too weak: the grazing angle measured in degrees should not exceed one-tenth of the wavelength measured in ångström.

On the contrary, the grazing incidence is required for smaller wavelengths—see Fig. 2.3. A spherical mirror must be quite long to accommodate the large area that is illuminated by a beam arriving at the grazing incidence. The corresponding large surface must be machined or mechanically bent to achieve the required curvature. This is not easy, since the curvature must have high accuracy. For hard X-rays with very small wavelengths, the grazing incidence angle becomes exceedingly small and the mirror surface extremely large, requiring the use of segmented mirrors.

Accurate focusing at the grazing incidence is complicated by the fact that a spherical mirror has two different focal points, one corresponding to focusing in the plane of incidence of the X-rays beam and the other in the plane perpendicular to it. Mirrors with non-spherical surfaces (elliptical, toroidal or parabolic) can eliminate this problem, but are more difficult to fabricate than spherical mirrors.

A simpler and less expensive solution to this problem is provided by the combination of two mirrors with cylindrical surfaces (obtained by bending or machining). The two mirrors independently focus the beam in the two perpendicular directions, producing a single final focal point for both directions. Such a device is called a 'Kirkpatrick–Baez' lens.

The difficulties in fabricating good curved-surface mirrors for X-rays are increased by the need to limit the surface roughness, which would otherwise weaken the reflection. This means that the average roughness must be very small compared with the wavelength λ (smaller, say, than $\approx \lambda/20$ at normal incidence). Considering that λ is very short for X-rays, this is a rather stringent requirement.

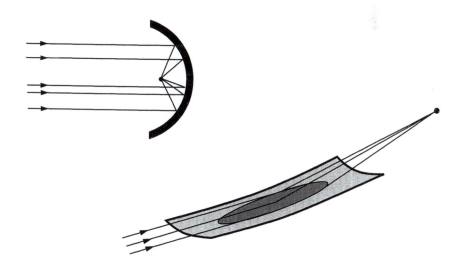

Fig. 2.3 Top: a curved focusing mirror for visible light. Because of the near-normal incidence, this type of mirror can only work for long wavelengths. Bottom: curved focusing mirror for short-wavelength X-rays. Note the large illuminated area.

Finally, all optical components along a beamline are affected by the problem of thermal stability. A synchrotron light beam carries a lot of power that can heat the optical devices and change their shape.

This problem is particularly severe for the first optical component placed at the beginning of each beamline, subject to the full power of the incoming beam. The fabrication of thermally stable optical components is an extremely difficult technical problem, requiring advanced designs, special materials and sophisticated cooling mechanisms.

The art of focusing X-rays can rely on other devices besides mirrors. Specialized optical components such as the Schwartzschild objectives and the Fresnel zone plates—discussed later—are quite advanced and can be rather effective.

2.1.1. Grating monochromators

Most applications of synchrotron light require only one wavelength—or, more precisely, a very narrow band of wavelengths—around a given value. Quite often, they also require the wavelength to be changed ('tuned') during an experiment, or to be scanned continuously over an extended spectral range.

We have seen (eqn 1.9 and Section 1.3) that the wavelength λ_L emitted by an undulator can be changed by modifying the undulator parameters—specifically, the B-field strength. The emitted bandwidth (eqn 1.11) is quite narrow, but not narrow enough for most applications of synchrotron light. Additional filtering is thus required to further reduce it. In the case of bending magnets, the emitted bandwidth is very large and spectral filtering is required in almost all cases.

This filtering is called 'monochromatization', and the corresponding devices are called 'monochromators'. The development of monochromators is a key problem in the synchrotron domain. Many types of instruments have been invented and successfully tested for different wavelength ranges.

Generally speaking, there exist two classes of X-ray monochromators: those based on *diffraction gratings* and those based on *crystals*. Grating monochromators work well for long-wavelength soft X-rays, down to $\lambda = 6$–12 Å. Crystal monochromators are used for shorter wavelengths.

A diffraction grating (Fig. 2.4) is a surface with a periodic array of many 'lines', i.e. geometric features of equal shape. A grating for X-rays must of course operate in reflection (whereas transmission gratings can be used for visible light).

Figure 2.4 illustrates schematically the working principle of a reflection diffraction grating. Two light beams reflected by equivalent points of two adjacent lines are 'in phase' if their overall path difference equals one wavelength, or an integer number of wavelengths.

As shown in Fig. 2.4c, the sum of the waves of two beams that are in phase has maximum amplitude, and therefore corresponds to maximum intensity. If the beams are slightly out of phase, then the combined wave amplitude decreases and so does the intensity. When the beams are totally out of phase (i.e. when their path difference equals $\lambda/2$ or an odd integer multiple of $\lambda/2$—see Fig. 2.4d), the combined amplitude and the intensity become zero.

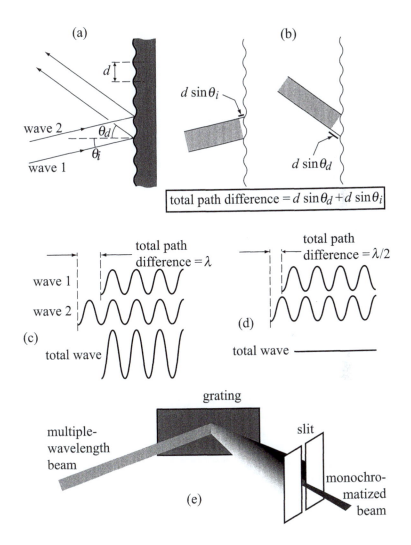

Fig. 2.4 Schematic explanation of a reflection diffraction grating. (a) The grating (side view) is a surface with a periodic array of features called 'lines' (d = period). Two light beams (waves 1 and 2) reflected by equivalent points of two adjacent lines travel along different paths. (b) Elementary trigonometry is used to calculate the total path difference. (c) If this path difference equals one wavelength (or an integer number of λs), then the superposition of the two waves produces maximum intensity. (d) If the path difference is one-half wavelength (or an odd multiple of $\lambda/2$), then the total wave has zero intensity. (e) The grating, combined with a slit, produces monochromatization.

Note that if two beams corresponding to adjacent lines are in phase, then all beams corresponding to all lines are in phase with each other. To understand this point consider, for example, two adjacent lines that correspond to a path difference λ. The path difference for any line and its next-nearest neighbor is 2λ, and in general the path difference for any two lines is an integer multiple of λ, so all the beams are in phase.

Trigonometry (Fig. 2.4b) quantifies the condition for maximum intensity in terms of the incidence and diffraction angles θ_i and θ_d. The path difference corresponding to adjacent lines is $d(\sin\theta_i + \sin\theta_d)$, and thus the condition for maximum-intensity is

$$d(\sin\theta_i + \sin\theta_d) = m\lambda , \tag{2.1}$$

where m is an integer number. According to eqn 2.1, a beam of given wavelength reaching the diffraction grating along a direction θ_i is diffracted with high intensity along different directions θ_d, each one corresponding to an integer value of m. The m-value defines the so-called 'order' of the corresponding diffracted beam.

How is a grating used to monochromatize X-rays? Imagine (Fig. 2.4e) a beam containing multiple wavelengths. For a given incidence angle θ_i, each wavelength is diffracted with maximum intensity at a specific set of angles θ_d, defined by eqn 2.1. By using a shield with a slit (Fig. 2.4e), it is possible to block the light in all directions except one of these θ_d-angles. This procedure filters out all wavelengths except the selected one.

This is of course a very simplified picture of monochromatization by a diffraction grating. More realistically, the primary beam is not a sum of discrete wavelengths, but a continuous distribution of wavelengths. The monochromatization process filters a narrow band of wavelengths out of this continuous distribution.

The width $\Delta\lambda$ of the monochromatized band of wavelengths is called the 'absolute resolution' of the monochromator. The quality of the monochromator increases when the value of $\Delta\lambda$ decreases. Quite often, however, the parameter used to characterize the monochromator performance is the 'resolving power' $\lambda/\Delta\lambda$. This parameter can be improved by using a grating with a large number of lines. In fact, a complete theoretical description of the grating mechanism would shows that the filtering effect of each line enhances that of all other lines. As a result, the resolving power is given by:

$$\lambda/\Delta\lambda = mN_L(\sin\theta_i + \sin\theta_d) , \tag{2.2}$$

where N_L is the total number of lines of the grating—or, more precisely, the total number of lines that are illuminated by the unmonochromatized X-ray beam.

Equation 2.2 shows that the resolving power improves if the number of illuminated lines increases. Unfortunately, fabricating a large-area grating with many lines is a difficult technical tasks that increases the cost.

The fabrication technology for gratings is quite sophisticated. For example, the lines must have a well-defined profile. The preferred profile (Fig. 2.5) is like a 'sawtooth'. The reason is simple: the actual diffracted intensity is the result of the combination of two mechanisms, diffraction and reflection. In order to optimize the intensity, we must simultaneously meet the diffraction condition of eqn 2.1 and optimize the reflection.

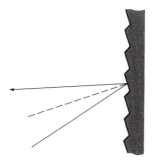

Fig. 2.5 Diffraction gratings with a sawtooth profile (side view) can be quite effective for specific wavelengths. The facets produce strong intensity in the direction of reflection. If this coincides with a maximum-intensity diffraction direction as defined by eqn 2.1, then the diffracted intensity is high. Equation 2.1 implies that this only occurs for a given wavelength, called the 'blaze wavelength' of the grating.

The 'sawtooth' profile of Fig. 2.5 provides a reasonable solution. The reflection is maximum if the large facets of the grating surface reflect the beam in the same direction as the maximum-intensity θ_d-angle of eqn 2.1. Unfortunately, this condition is valid for only one specific wavelength, called the 'blaze wavelength' of the grating. Thus, the grating is optimized for the blaze wavelength and for the spectral range close to it.

In a practical monochromator, the grating is combined with other components such as slits and mirrors to reach good performances—specifically, a high resolving power. In general, the beam is focused into an entrance slit that becomes the effective source point.

Then, it is spread to illuminate a large grating area and enhance the resolving power according to eqn 2.2. Finally, the beam is re-focused into an exit slit. Each monochromator design is based on a particular combination of grating, slits and mirrors, whose objective is a good compromise between focusing, monochromatization and the need to limit the intensity losses.

In fact, each optical component involves reflections that cause intensity losses; thus, their number should be limited. One can, in some cases, concentrate more than one function in the same component. For example, the grating can be fabricated on a non-flat (spherical or toroidal) surface to simultaneously obtain monochromatization and focusing. Such gratings, however, are difficult to fabricate and rather expensive.

The price of gratings can be reduced by fabricating masters and then using them to produce lower-cost replicas. The grating surface is typically coated with a film to enhance the grating performance. However, the high-intensity X-ray beam progressively degrades the coating and reduces its reflection. When the reflection becomes too weak, the grating must be re-coated or replaced.

Several different grating monochromator designs have been conceived over the past decades for synchrotron light experiments. Many of them are based on one of the oldest and most basic concepts in monochromator design: the 'Rowland circle', illustrated schematically in Fig. 2.6.

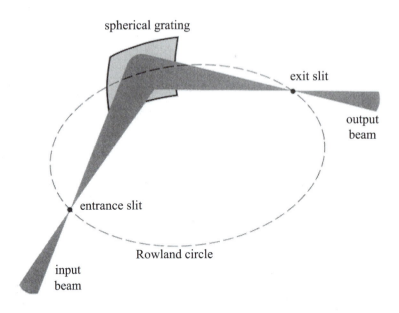

Fig. 2.6 Schematic of the Rowland-circle geometry for a spherical grating monochromator.

In the simplest case, the Rowland circle concept applies to a monochromator with three components: the entrance and exit slits and a focusing spherical grating. The entrance slit is the source point, and the grating must focus the diffracted beam into the exit slit. Theory shows that these requirements are satisfied if the grating center and the two slits are along a circle—the 'Rowland circle'. Unfortunately, the simplest Rowland-circle monochromator would have a moving exit slit that is difficult to use. Thus, more complicated designs are required.

Most of the advanced grating monochromator designs—based or not on the Rowland-circle concept—belong to broad classes. Each class includes variations of a 'parent' design, and some of the most important are the following:

- Normal-incidence monochromators (NIMs): because of the near-perpendicular incidence on the grating surface, such monochromators are only used in a rather limited range of long wavelengths. In this range, they achieve excellent levels of resolving power. Several types of NIM designs have been implemented, such as the Seya-Namioka geometry and the Wadsworth mounting (Fig. 2.7).
- Plane-grating monochromators (PGMs): the best performing instruments of this class are variations of the SX700 'parent' design (see Fig. 2.8).
- Spherical-grating monochromators (SGMs): after the introduction of the 'Dragon' design in the late 1980s (Fig. 2.9), many advanced instruments of this type have been based on variations of that concept.

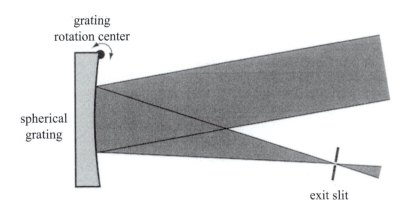

Fig. 2.7 An important type of normal-incidence monochromator design: the Wadsworth geometry. The characteristic elements are the spherical grating and the source point at an infinite distance from the grating, which eliminates the need for an entrance slit.

- Toroidal-grating monochromators (TGMs): the toroidal surface of the grating guarantees good focusing both parallel and perpendicular to the plane of incidence, thus removing the focusing problem affecting spherical gratings. However, the fabrication of gratings on toroidal surfaces is particularly delicate and expensive. Nevertheless, TGMs are widely used in most synchrotron facilities.
- Variable-line-spacing grating monochromators: the focusing action of the grating is obtained in this case by a variation of the periodicity of the grating lines, without requiring sophisticated curved surfaces.

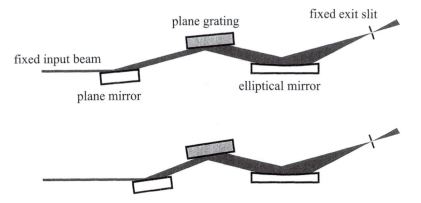

Fig. 2.8 The classic SX700 design for a high-performance plane grating monochromator geometry (see, for example, H. Petersen and H. Baumgarten, *Nucl. Instrum. Meth.* **173**, 191 (1980)). The monochromator is shown in two different positions yielding two different wavelengths. The characteristic components of the SX700 design are the rotating plane mirror at the entrance, the rotating plane grating, the focusing elliptical mirror and the fixed exit slit.

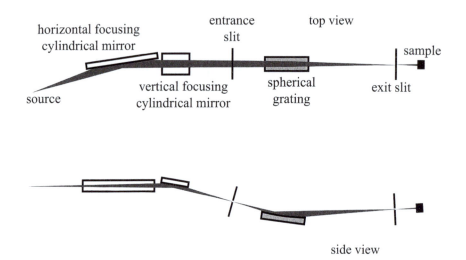

Fig. 2.9 Another classic grating monochromator design: the *Dragon* [see C. T. Chen and F. Sette, *Rev. Sci. Instrum.* **60**, 1616 (1989)], which is based on a spherical grating, on an entrance slit and an exit slit and on two cylindrical mirrors independently focusing the beam in perpendicular directions.

2.1.2. Crystal monochromators

The spectral range of hard X-rays is dominated, as we already mentioned, by crystal monochromators. The reason is quite clear if one considers the grating law of eqn 2.1. At grazing angles, $\sin\theta_i$ and $\sin\theta_d$ are both $\approx\lambda$ and the grating period d is $\approx m\lambda/2$. Therefore, even for high orders (relatively high values of the integer number m), d cannot be much larger than the wavelength λ. For short wavelengths, the period would become very small and fabricating the gratings is technically impossible.

On the other hand, the ordered atomic planes of a crystal automatically constitute the equivalent of a grating with a very short period. Another possible solution is to use artificial periodic structures such as multilayer thin films.

The mechanism exploited by crystal monochromators is 'crystal diffraction'. The basic law of crystal diffraction is the Bragg law, shown schematically in Fig. 2.10. Consider a beam of X-rays of wavelength λ that reaches a crystal, and assume that the distance between two adjacent atomic planes is d.

The two beams reflected by adjacent planes travel along different paths, and the path difference is $2d\sin\theta_d$. If this difference equals one wavelength, then the superposition of the two waves produces maximum intensity. This is the simplest form of the Bragg law for maximum-intensity crystal diffraction:

$$2d\sin\theta_d = \lambda \, . \tag{2.3}$$

The crystal thus acts as a 'natural grating', diffracting with maximum intensity a given wavelength in the direction identified by eqn 2.3. More realistically, however, we must consider that each crystal has many different families of crystal planes (see for example Fig. 2.10c). Therefore, eqn 2.3 applies to each of the corresponding d-values. The corresponding maximum-intensity directions (θ_d-angles) are called 'Bragg reflection'.

If the X-ray beam contains multiple wavelengths, then each wavelength produces a characteristic set of Bragg reflections. A shield with a slit can thus block the unwanted wavelengths and select only the desired one along one of its Bragg reflections. This is basically how a crystal monochromator works.

Suitable crystals for hard X-ray monochromators are selected primarily based on two criteria: their d-values and their crystal quality. As to the first criterion, the Bragg law of eqn 2.3 implies that the wavelength is smaller than $2d$. However, if λ is too small with respect to $2d$ then eqn 2.3 implies a very small θ_d-value—unsuitable for a practical monochromator. Therefore, the crystal must have $2d$-values that are neither smaller nor too much larger than the wavelengths of interest.

Good crystal quality is required for a high resolving power. In fact, a poor-quality crystal is formed by grains with different orientations. This produces a 'spread' of each Bragg reflection angle θ_d. In turn, the Bragg law of eqn 2.3 shows that a spread of θ_d corresponds to a spread in the wavelength, i.e. a limited resolution $\Delta\lambda$.

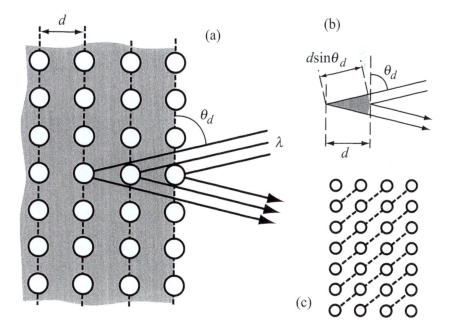

Fig. 2.10 (a) Schematic explanation of the Bragg law for crystal diffraction. (b) Derivation of eqn 2.3. (c) Another family of crystal planes for the same crystal of (a).

Similarly, a good resolving power requires a collimated beam. In fact, the angular spread of an insufficiently collimated beam negatively affects the wavelength resolution just like poor crystal quality. Therefore, a crystal monochromator works better with collimated synchrotron light than with a conventional source.

Besides the *d*-values and the crystal quality, other factors must be considered in selecting a crystal for a monochromator. The crystal, for example, must be thermally stable and capable of being exposed to synchrotron light with limited radiation damage. The actual performance of a crystal monochromator must be estimated taking into account possible multiple scattering of the X-ray photons rather than only single-step Bragg scattering.

The overall quality of a crystal monochromator is characterized by the so-called 'rocking curve'. This is the plot of the Bragg-reflected intensity as a function of the θ_d-angle. In an ideal case (perfect crystal, perfectly collimated beam), the rocking curve would be an infinitely narrow peak centered at the Bragg angle defined by eqn 2.3. In reality, the finite beam collimation and limited crystal quality result in a finite peak width.

Similar to grating monochromators, many different crystal monochromator designs have been invented in the past decades. Focusing is again a problem, and curved crystals are often used to obtain both monochromatization and focusing.

The curvature can be achieved by mechanically bending the crystal. This is relatively easy in one direction, which gives focusing in the same direction, but more difficult for two perpendicular directions.

An effective way to improve the resolving power is to use Bragg reflections by two different crystals (Fig. 2.11). This approach is used by the so-called 'double-crystal monochromators'. In some cases, more than two crystals are used.

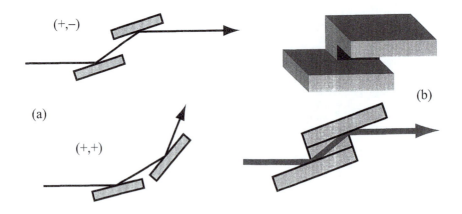

Fig. 2.11 (a) Two different geometries for double-crystal monochromators, labeled as (+,−) and (+,+). (b) Idealized picture of a channel-cut crystal (top) and the corresponding monochromator geometry.

Some double-crystal monochromators use, instead of two independent crystals, two parts of the same, large-quality crystal, called a 'channel-cut' crystal (see Fig. 2.11). This approach guarantees perfect alignment of the two crystal surfaces, but decreases the flexibility of the monochromator operation.

When two independent crystals are used, the double-crystal monochromator design must reconcile good performance (spectral tunability, spectral resolution, high brightness, stability, etc.) with simplicity and a reasonably limited cost. The design must specifically provide simple but accurate control of the positions of the two crystals under specific constraints (e.g. a fixed exit beam direction).

Over the years, many designs have been proposed and implemented to solve this problem. Quite noteworthy are the variations of the parent 'boomerang' design. This name originates from a right-angle boomerang-shaped linkage between the two crystals. To change the wavelength, the second crystal must be both translated and rotated while remaining parallel and at a fixed offset with respect to the first crystal. Figure 2.12 shows schematically how these conditions are guaranteed by the 'boomerang' linkage.

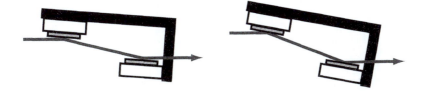

Fig. 2.12 Scheme (for two positions) of the 'boomerang' geometry for double-crystal monochromators, based on a boomerang-shaped linkage (in black) between the crystals. For the original design see: P. L. Cowan, J. B. Hastings, T. Jach and J. P. Kirkland, *Nucl. Instrum. Methods* **208**, 349 (1983).

2.2. Vacuum systems

Stringent vacuum requirements constitute a key technical difficulty in using synchrotron sources. We have seen that the storage ring must be under 'ultrahigh vacuum' (UHV) to avoid collisions between the circulating electrons and residual gas particles. UHV corresponds to pressures lower than 10^{-7}–10^{-8} Pascal [1 Pascal (Pa) is a pressure of 1 Newton per square meter, corresponding to approximately 10^{-5} atmospheres].

Furthermore, many synchrotron beamlines must also be kept under UHV. This is due to the absorption properties of X-rays by materials. Specifically, soft X-rays are absorbed by all solid materials.

Therefore, a soft X-ray beamline must be connected without any window to the storage ring, and must be kept at a pressure compatible with the ring. The vacuum requirements are less stringent for hard X-rays at lower photon energies: the beamline can be separated from the ring by a partly transparent beryllium window and kept under more moderate vacuum—or even no vacuum at all.

The UHV technology required to achieve and maintain pressures below 10^{-7} Pa is quite sophisticated but well developed and commercially available. For such pressures, the most widely used pumps are ion pumps. They can only work at pressures lower than 10^{-2}–10^{-3} Pa; therefore, they must be used in conjunction with 'roughing pumps' that bring the pressure from atmospheric level to 10^{-2}–10^{-3} Pa or lower.

Let us briefly elaborate on the vacuum requirements. Without UHV in the ring, the circulating electrons would be scattered quickly by the residual gas particles and leave the electron beam, whose short lifetime would be unsuitable for practical experiments.

This phenomenon can be visualized as a bullet (an electron) fired against a clump of trees (the residual gas particles). Can the bullet pass the clump without being stopped by one of the trees?

The answer is that the probability of passing depends on the length of the clump and on the density of trees. A low density leaves much free space for the bullet to pass the clump without being stopped. In the case of residual gas particles in a ring, the density is determined by the residual gas pressure p.

Specifically, elementary physics shows that the distance L ('mean free path') along which the electron-bullet can travel before being scattered by a residual gas particle is given by

$$L = \text{constant}/p \ . \tag{2.4}$$

Therefore, L increases as p decreases.

If the ring is at atmospheric pressure, then L is so short that an electron is not even able to travel once around the ring. The pressure in the ring must thus be lowered below atmospheric pressure. By how much? Realistically, most experiments require an electron beam with a lifetime of the order of 10 h to one day. This is possible if the pressure is lower than 10^{-7}–10^{-8} Pa—thus, the ring must indeed be kept under UHV.

What are, on the other hand, the vacuum requirements for a beamline? Consider first hard X-rays: if the wavelengths is below ≈ 4 Å, an X-ray beam can travel along several meters in a gas at atmospheric pressure without being substantially attenuated. At shorter wavelengths this distance is even higher: for example, at 1.2 Å it exceeds 1 km. Therefore, a vacuum is not strictly required for a hard X-ray beamline.

As the wavelength increases reaching the soft X-ray range, the absorption becomes more significant and vacuum becomes necessary. Roughly speaking, a beamline under vacuum is not required for hard X-rays of wavelength < 3–4 Å. In some cases, however, specific experiments require a hard X-ray beamline to be kept under moderate vacuum with a conventional pumping system rather than with the sophisticated technology for UHV.

When the beamline is not under UHV, it must be separated from the storage ring that works at lower pressure. This is accomplished by inserting a thin window made of beryllium.

Why beryllium? The reason is that this element is a weak absorber of hard X-rays. The main physical mechanism for absorption is the excitation of core electrons, discussed in detail in the next chapter. We can anticipate here one important point: 'light' elements (elements with a low atomic number) absorb X-rays much less than 'heavy' elements.

Beryllium is a very 'light' element (atomic number 4) and therefore its X-ray absorption is weak. At the same time, it is quite suitable for thin but mechanically strong windows.

For example, a 0.5 mm thick beryllium plate absorbs only ≈5% of a $\lambda = 1.2$ Å X-ray beam (and much less at lower wavelengths). Note, however, that beryllium windows cannot be used for soft X-rays: for example, even a very thin sheet would completely absorb a beam with a 12 Å wavelength.

The vacuum system of a synchrotron source has a special characteristic: it is particularly vulnerable to accidents. During an experiment, a soft X-ray beamline is directly connected to the ring without any window. A sudden vacuum leak in the beamline would affect the vacuum in the entire ring and also in all other windowless beamlines that are open to the ring at the time of the accident.

The problem is somewhat less severe for hard X-ray beamlines. However, a sudden pressure change can damage or break the thin beryllium windows causing again problems for the entire ring.

This vulnerability of the vacuum system requires special precautions. Fast valves are often inserted between the central ring and each beamline, and are rapidly and automatically closed in the event of a sudden pressure increase. More conventional valves separate each beamline and the ring and must remain closed when the beamline is not in operation (radiation safety also requires a 'radiation shutter' in front of the beamline to be closed).

The use of some types of vacuum equipment is forbidden. This applies for example to oil diffusion pumps because of the risk of oil contamination that could affect the entire vacuum system of the synchrotron facility.

After a complicated pumping process, the final path towards UHV is the elimination of gas molecules trapped by the walls of the vacuum chambers. Such molecules would otherwise be slowly released and make it impossible to reach UHV in a reasonably short time. The procedure to solve this problem by eliminating the gas molecules from the walls is called 'bake out'.

This procedure consists of keeping the entire vacuum system at high temperature (at least above 100 C) for a day or two. A successful 'bake out' requires a reasonable homogeneous temperature. This is achieved by wrapping the vacuum system in aluminum foil—a fact that invariably stimulates the attention and questions of first-time visitors, from high-level political leaders to schoolchildren.

Even after the bake-out, gas molecules remain in the wall. The synchrotron light beam is very powerful and forces their release much more effectively than the heating of the wall. Thus, the synchrotron beam performs the real bake-out for the portions of the vacuum chamber that are under illumination. This process can last for days or weeks, aggravating the consequences of vacuum accidents affecting the main ring.

2.3. Experimental chambers

Each beamline ends with one or several experimental chambers, equipped with several instruments. There exists a wide variety of beamline instruments and experimental configurations, with considerable differences between different synchrotron sources.

Potential users interested in specific instrumentation should directly consult the documentation provided by each synchrotron facility. They can specifically verify what kind of instrumentation is available for each beamline at a given site. This information is typically available through Internet (see Section 2.4).

Later chapters in this book provide details about instruments used in specific applications. We briefly discuss here two classes of devices that are widely present in synchrotron beamlines: photon detectors and electron analyzers.

2.3.1. Photon detectors

The oldest and still (considering medical radiology) most widely used type of X-ray detector is the photographic plate. The local color level in a developed photographic plate is determined by chemical reactions stimulated by the absorbed photons, and depends on the local absorbed photon beam intensity.

Photographic plates are a well-established technology with many advantages—but also serious drawbacks. Their sensitivity is limited at low exposure levels, and their response may not be linear (this means that the recorded intensity in some cases is not proportional to the number of detected photons).

Another problem is spatial resolution. A photographic plate does provide the spatial resolution required for many synchrotron-based techniques (specifically for radiology). However, the resolution is limited by the grain size of the photographic emulsion.

Furthermore, it is difficult to compare results of different photographic plates. In fact, the intensity (gray) scale of each plate depends on the developing process and absolute normalization encounters serious obstacles.

Finally, photographic plates cannot easily handle the problems of data storage and processing. Synchrotron light techniques can produce a huge flow of data that often require fast and sophisticated processing. This requires high-performance computers, large storage capacity and fast transmission. The processed data must be in digital form: the photographic plates must therefore be converted to digital files with a somewhat cumbersome process.

Such problems cause a progressive shift away from photographic plates and towards detectors that directly produce electronic data suitable for digitalization. All types of such detectors convert the detected photon (or photons) into an electronic signal, for example a current. Each class of detectors, however, uses a different mechanism for the conversion.

2.3.1.1. Ionization chambers

In detectors of this type, X-rays pass through a window and reach a gas-filled chamber. The X-ray photons are absorbed and cause the ionization of gas molecules, i.e. the molecules lose negatively charged electrons and become positively charged.

The ionization chamber has a positively biased wire (anode) and a negatively biased cathode (the gas container). The photon-ionized molecules and the corresponding free electrons are separated by the voltage difference between the anode and the cathode. As the charges reach the cathode and the anode, they create a current in the external detector circuit. The X-rays are detected by measuring this current.

With a large voltage difference between the anode and cathode, the free electrons are strongly accelerated and cause secondary ionization events on their own. Each absorbed X-ray photon thus produces an avalanche of ionization events. The current signal is thus amplified and easier to measure.

2.3.1.2. Proportional counters

This type of detector is a variation of the ionization chamber scheme. With a very large avalanche amplification, each single absorbed photon produces a measurable current pulse in the external circuit. The number of detected photons is thus obtained not by measuring the total current but by electronically counting the individual photon-generated pulses. The detector is no longer called an ionization chamber but a 'proportional counter'.

The peak value of each photon-generated current pulse depends on the energy of the X-ray photon, since this energy determines the strength of the avalanche mechanism. The detected photon energy can thus be derived from the pulse height.

A proportional counter can achieve spatial resolution with a rather simple approach. Imagine the arrival at a given point on the wire of the charge produced by a photon-induced ionization avalanche. This will produce two separate pulses traveling towards the two ends of the wire.

The two pulses reach the two ends of the wire at different times. From the time difference, the point of origin can be identified: this is the photon detection point along the wire.

2.3.1.3. Scintillation counters and electro-optical detectors

These detectors are based on the phenomenon called 'scintillation' or, more accurately, 'photon-stimulated luminescence'. When a material absorbs an X-ray photon, the photon energy is taken by one of its electrons that goes temporarily into an excited state. As the electron falls back into its ground state, it must release the excess energy. This can cause the emission of a photon, i.e. photon-stimulated luminescence.

Each absorbed photon can be revealed by detecting the burst of light produced by this mechanism, and the total number of absorbed photons can be measured by counting the light pulses. This approach can be implemented using a variety of materials where photon-stimulated luminescence occurs, including solids, liquids and gases.

In electron-optical detectors, the X-ray beam produces an image by stimulating luminescence from a solid surface coated with a phosphor material such as ZnS. The image is typically very weak, and its intensity must be enhanced with an amplification device. The amplified image is then detected, for example with a television camera.

2.3.1.4. Semiconductor detectors

The operation of these devices is to some extent similar to that of proportional counters: the absorbed photons create pairs of positive and negative charges that can produce an electric signal in the external circuit. However, instead of creating electron–ion pairs in a gas, the absorbed photon produces pairs of 'free electrons' and 'free holes' in a semiconducting material, see Fig. 2.13a.

The 'free holes' are charge carriers similar to the free electrons in a conductor, except that their charge is not negative but positive. Together with the free electrons they contribute to the total current that is present in a semiconductor with an external voltage bias.

By creating a series of electron–hole pairs, each absorbed photon produces a current pulse in the external circuit. The pulse intensity is proportional to the number of electron–hole pairs created by the photon, which in turn increases with the photon energy.

Therefore, by measuring the pulse intensity we can derive the energy of the corresponding photon. The conversion of electron–hole pairs into an electronic pulse is performed by a microdevice built on the surface of the semiconducting material.

Figure 2.13b shows schematically the cross-section of a semiconductor detector with spatial resolution, in which many detecting microdevices form a regular array. By individually 'reading' the microdevices, the control instruments of the detector find where each photon is absorbed, thus achieving spatial resolution.

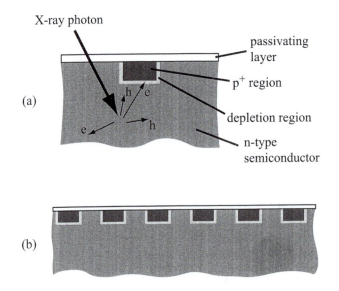

Fig. 2.13 Simplified schematic of a semiconductor X-ray detector. (a) The absorption of a photon by a semiconducting material produces electron–hole (e–h) pairs. Such pairs can be collected by a microdevice at the semiconductor surface and produce a pulse in an external electronic circuit (not shown). Here, the microdevice is a PIN junction produced by MOS (metal–oxide–semiconductor) technology and protected by a silicon dioxide passivation layer. The PIN junction is essentially a microcapacitor kept at constant external bias. The arrival of free carriers in the isolating (depletion) layer produces a discharge of the microcapacitor and an electronic pulse. (b) Scheme an array of microdetectors to reveal photons with high lateral resolution. A real array detector contains of course a much larger number of microdetectors.

The spatial resolution level of semiconductor detectors is excellent. X-ray photons are more strongly absorbed by semiconducting solids than by gases, thus producing charge pairs in a more localized region. Furthermore, the detecting microdevices can be made extremely small with the sophisticated fabrication techniques of microelectronics.

Several types of microdevices can be used as detecting elements, such as asymmetric p–n junctions and PIN structures. As far as materials are concerned, most semiconductor detectors are fabricated with doped elemental semiconductors such as Ge(Li) or Si(Li).

2.3.1.5. CCD (charge coupled device) detectors

This is an important type of semiconductor device, not only for synchrotron beamlines but mainly for consumer items such as camcorders and digital cameras. How does a CCD work?

When detecting a photon, a microdetector of the CCD becomes charged. Assume that this charge is transferred from the microdetector to its nearest neighbor, and then to the next neighbor, and so on. In this way, the photon-generated pulses in the entire array reach the array edge where they are 'read' by the control electronics.

How is the charge transfer between microdetectors triggered? The mechanism is schematically illustrated in Fig. 2.14 (top) using as a model an array of water containers with pistons. Water can move from one container to the next through a hole. The synchronized movement of the pistons produces the translation of the water buckets— as shown in Fig. 2.14 by comparing the situation at two different times.

This model, although very simple, clearly shows that the individual water buckets are translated without interfering with each other. Therefore, at the end of the array they can be detected as individual 'pulses'.

An actual CCD detector consists of MOS microelements. Its operation cycle starts with an 'accumulation' phase: a voltage bias transforms the MOS elements into localized charge traps. When a photon is absorbed, it creates a 'bucket' of charge that is captured by one of the MOS elements. After the accumulation phase ends the charge in each bucket is individually moved to the end of the MOS array. This is done by dynamically and synchronously changing the bias voltages of the individual MOS elements.

Note that the charge transfer between adjacent microelements is only possible if the microelements are very close to each other. Thus, the fabrication of CCD detectors requires the miniaturization techniques of microelectronics. The bottom part of Fig. 2.14 shows a simple version of an MOS CCD array. Real CCDs for X-ray detection are much more complicated.

The CCDs are excellent image detectors. This function is achieved in the following way. The charge stored by a given MOS element during the accumulation phase reaches the end of the array during the 'read' phase and is detected at a time that depends on the element position. The control instruments thus capture a sequence of pulses whose individual intensity and timing carry information on the detected X-ray beam intensity at each point of the array. This is the digital equivalent of a line image. A similar approach is used for two-dimensional images by sequentially reading the pulses produced by all the 'lines' of a two-dimensional array.

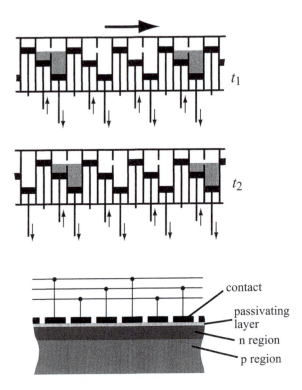

Fig. 2.14 Simplified schematic of a charge coupled device (CCD). Top: model based on an array of water containers and pistons. The translation of water buckets (in gray) is made evident by comparing the situation at two different times ($t_2 > t_1$). Bottom: a simple type of CCD detector array fabricated with MOS technology.

Because of their superior efficiency and of their rapidly improving spatial resolution, CCD X-ray detectors are finding expanding applications in diverse fields such as crystallography and digital radiology. The spatial resolution (reaching a level of a few microns) is already adequate for important applications such as crystallography.

In fact, the key problem is not so much the spatial resolution but the fabrication of large-area detectors. The corresponding two-dimensional arrays must have an overall good quality and a limited fraction of malfunctioning microelements. Both of these characteristics are difficult to achieve over large areas. In some cases, an X-ray image is first demagnified and then detected, thereby degrading the resolution (within acceptable limits) but effectively increasing the detected image area.

2.3.1.6. Channel multipliers and channel plates

The operating principle of a channel multiplier detector is explained schematically in Fig. 2.15 (top). An absorbed X-ray photon extracts (e.g. by photoelectric effect) an electron from the inside surface of the 'channel'.

The voltage difference between the two channel ends accelerates the emitted electron. If the electron hits the channel surface, then it has sufficient energy to cause the extraction of more electrons. This leads to an avalanche mechanism: each absorbed photon produces a bunch of electrons at the end of the channel and an electronic pulse in the control circuit (nor shown).

Channel multipliers are good detectors for soft X-rays and for ultraviolet photons. Lateral resolution can be achieved in this case by packing together very many small channel multipliers in a 'channel plate'—see Fig. 2.15 (bottom).

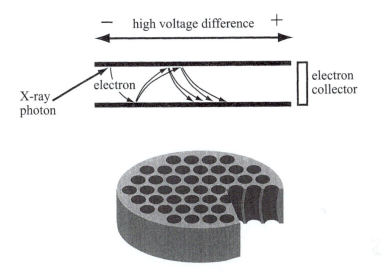

Fig. 2.15 Top: schematic of a channel multiplier detector. Bottom: section view of a channel plate, formed by an array of channel multipliers with a shape that enhances the interaction of photons or electrons with the channel walls.

2.3.2. Electron analyzers

Many synchrotron-based experimental techniques for chemistry and for the life science— e.g. photoelectron spectromicroscopy—require the detection and analysis of electrons. Electron detection can be based on channel multipliers and channel plates (Fig. 2.15): the avalanche mechanism in this case is triggered by the arrival of an electron rather than by the absorption of a photon.

As to the analysis of the electron properties, the two key objectives are measuring the energy and the direction of the electron movement. The simplest way to analyze electron energies is to use a retarding potential (illustrated schematically in Fig. 2.16).

Only electrons with energy larger than the retarding potential can overcome the potential barrier and be detected. By scanning the retarding potential ($-V_1$ in Fig. 2.16) we can thus obtain the electron distribution as a function of energy.

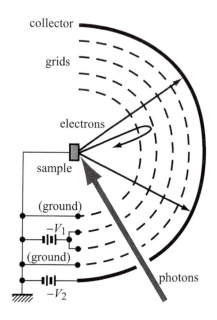

Fig. 2.16 Schematic of a four-grid retarding-potential electron analyzer, used in this case to detect and analyze electrons emitted by a sample under photon bombardment. The retarding potential between the inner grid and the two central grids ($-V_1$) makes it impossible for electrons of energy smaller than eV_1 to pass and reach the external collector. The multiple grids improve the energy resolution, and the additional voltage drop $-V_2$ eliminates stray electrons.

Retarding-potential analyzers can achieve good energy resolution by using multiple grids, as shown in Fig. 2.16. However, they cannot measure the direction of the electron motion.

This becomes possible by using angle-resolved analyzers. Important types of angle-resolved devices are the cylindrical mirror analyzers (single or doubles-pass, with a pinhole direction filter) and the spherical-sector analyzers (see Fig. 2.17). Each one of these devices uses an electric field to modify the trajectory of the electrons. The modified trajectory depends both on the initial electron energy and on the initial electron direction.

After moving through the electric field and before being detected the electrons are filtered by a slit system. This eliminates all the electrons except those with a given modified trajectory. The detection is thus limited to the corresponding initial electron energy (called the 'pass energy') and initial direction. An adjustable pre-retarding electric field is often used to bring the initial electron energy to the pass energy value. By scanning the magnitude of the pre-retarding field, we can thus measure the energy distribution of the electrons for a given initial direction, performing at the same time direction and energy analysis.

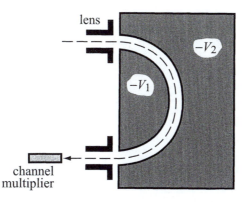

Fig. 2.17 Schematic of spherical-sector angle-resolved electron energy analyzer. The electric field produced by the voltages V_1 and V_2 modifies the electron trajectory. Only electrons of a given initial energy and initial direction can follow the dashed trajectory and be detected. A pre-retarding field can be incorporated in the entrance lens and used for energy scanning.

Substantial improvements brought the performances of angle-resolved analyzers to excellent levels. For example, the energy resolution in analyzing electron energies of 10–20 eV reaches values of a few millielectronvolts, much better than the typical 0.1–0.2 eV values of the early 1990s.

2.4. Becoming a synchrotron light user

After becoming synchrotron users, scientists can appreciate the excellent new opportunities provided for their research by a synchrotron facility. Quite often, however, the first steps are not easy: becoming a synchrotron user requires overcoming practical as well as psychological obstacles.

Most new users are from 'small science': they normally conduct experiments in laboratories of limited size involving a few people. They are not familiar with large-scale centralized research facilities. Understandably, they are sometimes intimidated by the unknown environment and concerned about technical and logistic problems.

Most of such problems, however, are more psychological than real. The pioneering hard times of synchrotron research are (fortunately) past history. Today, synchrotron facilities are well organized to serve different types of users, including occasional users and newcomers with no previous experience. They share with all of their customers the same objective: producing successful experiments.

Synchrotron users fall into different categories depending on the scope of their synchrotron-based activities. Some groups are permanently or semi-permanently stationed at a synchrotron facility, where they develop and operate their own equipment. These users are expert and therefore largely independent of the support provided by the facility.

Most groups, however, use synchrotron light with a much more limited scope. Synchrotron-based experiments constitute only a portion of their research activities and do not justify nor require the development of special equipment and the presence *in situ* of permanent staff. It must be stressed that the research of such 'occasional' users is not inferior to that of permanent users. They in fact systematically produce some of the most important synchrotron-based experiments.

Synchrotron facilities provide excellent support for their 'occasional' users, in particular during the first steps. Their entrance in the world of synchrotron light is easier than most scientists think.

First of all, the access to synchrotron beamtime is free of charge in almost all facilities: the only condition is the non-proprietary nature of the research, whose results must be destined for publication. The beamtime is allocated through open competition based on merit and peer review. This may cause some problems for new users not expert in synchrotron techniques. However, the facility usually provides assistance for newcomers during the competition process.

After obtaining beamtime, the 'occasional' users are assisted at each step of their experiments by the professional staff of the facility. In many cases, all they have to do is to bring their samples and their ideas.

As a general trend, synchrotron facilities are becoming increasingly 'user-friendly' to enhance their productivity and justify their financial support. New users quickly realize that the seemingly complicated beamlines are, in most cases, quite easy to operate. They thus become gradually independent and require less help by the facility staff.

Additional support is often provided by other users. In fact, groups working at a synchrotron facility often retain their 'small science' spirit and openness. Note that a synchrotron source is *not* reserved for a few complicated experiments. On the contrary, it supports a very large number of small-scale experiments, quite similar to those found in small laboratories. A synchrotron source, in other words, is not 'big science', but a big collection of 'small science' experiments.

What is the best way to become a synchrotron user? The first step is to obtain information. Today, the most effective vehicle for information is the Internet. Each facility maintains a sophisticated Internet site and extensive lists of synchrotron homepages can be found at the following addresses:

http://www.elettra.trieste.it/sites/synchrotrons.html
http://www-ssrl.slac.stanford.edu/sr_sources.html
http://www1.psi.ch/www_sls_hn/sls_other_laboratories.htmlx

whereas this site provide links to most of the free electron laser (FEL) facilities:

http://sbfel3.ucsb.edu/www/vl_fel.html

In its homepage, each facility provides a simple explanation of its mission and organization, plus detailed information on specific equipment available for users. The site also explains the procedure to apply for beamtime, the competition rules and the corresponding deadlines.

The Internet is in general a very valuable source of information and resources for synchrotron-based science and technology. We note in particular the following addresses because of their general interest:

- http://www-cxro.lbl.gov/—this is the homepage of the Center for X-ray Optics at the Lawrence Berkeley Laboratory. The site contains a wealth of information on X-rays, including extensive databanks for optical constants and access to the well-known *X-ray Data Booklet*.
- http://physics.nist.gov/PhysRefData/XrayMassCoef/—this is part of the NIST (National Institute of Standards and Technology) site, and is dedicated to a databank on X-ray absorption.
- http://www.csrri.iit.edu/—homepage of the Center for Synchrotron Radiation Research and Instrumentation, full of very valuable information on X-rays and on the use of synchrotrons.
- http://www.bmsc.washington.edu/scatter/periodic-table.html—another good source of information about X-ray constants and properties, offered by the Biomolecular Structure Center of the University of Washington.
- http://nslsweb.nsls.bnl.gov/infrared/science/srn_bio.htm—excellent source of information on the infrared applications of synchrotrons, housed at the Brookhaven National Laboratory.

Finally, it must be stressed that synchrotron light is not a private club reserved for 'rich' users. For most users, the costs need not to constitute a problem. First of all, we have already seen that the facility provides the beamtime and the corresponding support free of charge.

Even so, many users might encounter difficulties in financing their travel and lodging expenses. Some of the major facilities—notably the European Synchrotron Radiation Facility (ESRF) in Grenoble—do provide partial or total travel and lodging support for their users. This is also the standard policy for the twelve member facilities of the European Commission Round Table for Synchrotron Radiation and Free Electron Lasers. These facilities obtain special European Commission contracts to finance the user access costs. Finally, almost all facilities provide low-cost housing.

In summary, the first steps towards the use of synchrotron light need not to be intimidating, and can be quite inexpensive. If a scientist or a technologist suspects that synchrotron light might help his/her research, then the best suggestion to him/her is: stop wondering and try it!

3. Applications of synchrotron light

A synchrotron source provides the photons used for a wide variety of techniques and for an even wider variety of applications. Photons interact with atoms, molecules, solids and complex biological systems in many different ways—see Fig. 3.1. Each one of the interaction mechanisms constitutes the basis of several experimental techniques.

For example, synchrotron photons can be absorbed by atoms or molecules in a gas, and their energy can be used to:

- Put an atom or a molecule into an excited state. The decay from the excited state can cause the emission of secondary photons (fluorescence). The corresponding experimental techniques are *absorption spectroscopy, molecular EXAFS* (extended X-ray absorption fine structure) and *fluorescence spectroscopy.*
- The absorbed photon energy can also be used to extract electrons from the atom or molecule. The corresponding techniques are again *absorption spectroscopy* and *molecular EXAFS*, plus *gas-phase photoelectron spectroscopy* and *Auger-electron spectroscopy.*
- The absorbed energy can break a molecule to produce smaller fragments. The corresponding technique is *molecular fragmentation spectroscopy.*

Instead of being absorbed, the X-ray photons can be scattered, either elastically (i.e. without loss of energy) or inelastically. These mechanisms constitute the basis of many of *X-ray scattering* techniques.

We can now move from atoms and molecules to solids and complex biological systems. The photon absorption in such systems provides the basis for *absorption spectroscopy* and *solid-state EXAFS*, plus *X-ray imaging* (and *radiology*), *X-ray microscopy* and *fluorescence analysis and microanalysis.*

Photons interacting with solids can also be scattered and diffracted. These mechanisms are widely used by *crystallography* (in particular, *powder diffraction* and *protein crystallography*) and by other techniques such as *topography* and *small-angle X-ray scattering*. Alternatively, the photons can be elastically or inelastically scattered, leading to a variety of *scattering techniques* and in particular to *small-angle X-ray scattering.*

The absorption of photons can stimulate the emission of electrons from solids, exploited for *solid-state photoelectron spectroscopy, photoelectron spectromicroscopy* and *Auger spectroscopy*. Finally, the photons can stimulate the desorption of atoms and molecules or other aggregates, providing the basis for *photon-stimulated desorption spectroscopy.*

This is, of course, only a very partial list of synchrotron-based techniques. In the different sections of this chapter we discuss, in very simple terms, a subset of widely used synchrotron techniques that are specifically important for chemistry, biology and medical applications.

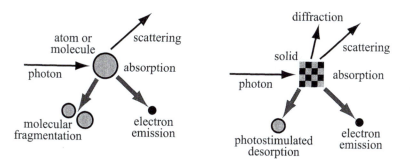

Fig. 3.1 An ultraviolet or X-ray photon can interact in many different ways with an atom, a molecule or a solid. Each one of these mechanisms is exploited by several synchrotron-based experimental techniques.

3.1. X-ray imaging

This is the oldest applications of X-rays: Röntgen inaugurated radiology only a few hours after discovering them. A few years later, radiology was already a flourishing domain with several industrial spinoffs. However, its development was negatively affected for many decades by the lack of advanced X-ray sources.

The technical progress in X-ray production was quite limited until the advent of synchotron light more than seventy years after Röntgen's discovery. Furthermore, the use in radiology of synchrotron X-ray sources has been limited: even now, the full power of advanced synchrotron sources is only marginally exploited.

This is totally unjustified, since the few existing radiological applications of bright and coherent synchrotron sources produce rather spectacular results. Figure 3.2, for example, shows that the coherence of synchrotron light can lead to an amazing enhancement of the radiological contrast, while decreasing the X-ray dose.

In order to discuss results of this kind, we must first analyze the basic absorption mechanism of X-rays by solids (or biological materials). Our description is qualitative, whereas Inset F presents a more detailed theoretical treatment.

3.1.1. X-ray absorption by solids and biological materials

We will use two subsequent approaches in order to describe the interaction of X-rays with a solid material. The first approach is empirical and ignores the actual microscopic mechanisms. In thus provides valuable tools for treating X-ray absorption, but it cannot fully justify how the absorption takes place. The second approach treats instead the microscopic interaction mechanisms at the atomic level.

3.1.1.1. An empirical approach

Consider (Fig. 3.3) an electromagnetic wave entering a solid material. At the external surface, the wave can be reflected and/or scattered. In the case of X-rays, however, we are more interested in what happens *after* they pass the surface and enter the solid.

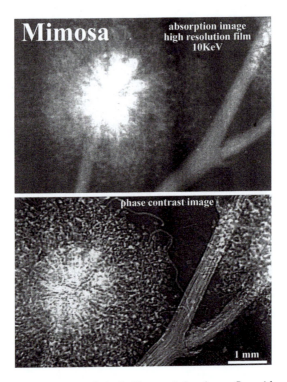

Fig. 3.2 Direct comparison of two radiological images (of a mimosa flower) based on conventional absorption contrast (top) and on phase contrast (bottom), which requires a coherent synchrotron X-ray source. Details of the results can be found in: F. Arfelli, M. Assante, V. Bonvicini, A. Bravin, G. Cantatore, E. Castelli, L. dalla Palma, M. de Michiel, R. Longo, A. Olivo, S. Pani, D. Pontoni, P. Poropat, M. Prest, A. Rashevsky, G. Tromba, A. Vacchi, E. Vallazza and F. Zanconati, *Phys. Med. Biol.* **43**, 2845 (1998). See also: G. Margaritondo, *Phys. World* **11**, 28 (1998).

3.1.1.2. The absorption coefficient

Figure 3.3 (top) illustrates the definition of α, based on the fact that the intensity I of the wave beam decreases as the distance x from the surface increases. Such a decrease follows an exponential law:

$$I(x) = I_0 \exp(-\alpha x) . \tag{3.1}$$

Note that $I(x)$ is a decreasing function since its argument $(-\alpha x)$ is negative. Over a certain distance x, the decrease is more rapid if the magnitude of the absorption coefficient α is larger. Figure 3.3 (bottom) illustrates how the curve corresponding to eqn 3.1 changes for different values of α.

The plots of Fig. 3.3 suggest that most of the wave intensity decrease occurs over a distance of the order of $1/\alpha$. This quantity—called the 'attenuation length'—can thus be loosely interpreted as the typical path traveled by the wave before being absorbed.

Finally, when $x = 0$ (the surface of the solid) the argument $(-\alpha x)$ of the exponential in Eq. 3.1 is also zero, and $I(x) = I_0$. Therefore, the quantity I_0 is the initial intensity of the wave beam when it passes the external surface.

The value of the absorption coefficient changes of course from material to material, and also as a function of the wavelength. For visible light, α is almost zero in the case of a transparent glass. However, a colored glass selectively absorbs certain wavelengths. Thus, its corresponding absorption coefficient is large for those wavelengths.

Consider now wavelengths in the domain of ultraviolet light or X-rays. We already mentioned in the previous chapter that soft X-rays (and most ultraviolet photons) are strongly absorbed by all solid materials. On the contrary, hard X-rays can pass through thick materials without much absorption. These facts can be described in terms of the absorption coefficient.

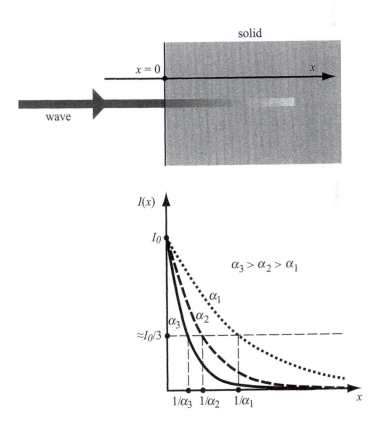

Fig. 3.3 Definition of the absorption coefficient α. Top: as a wave passes the surface of a solid and travels into the bulk, its intensity I is progressively attenuated according to eqn 3.1. Bottom: the intensity decay rate depends on the absorption coefficient α, and is higher if the absorption coefficient value is larger. Roughly speaking, the initial intensity I_0 decays by approximately two-thirds over a distance $1/\alpha$, which is called the 'attenuation length'.

Figure 3.4 shows for example the plot of α as a function of the wavelength for solid polypropylene (C_3H_6). Consider an ultraviolet beam with wavelength $\lambda = 300$ Å: the corresponding value of $1/\alpha$ is only $\approx 10^{-7}$ m or one-tenth of a micron. Clearly, this beam is very quickly absorbed and cannot pass through the material. In fact, its penetration depth beyond the surface is extremely short.

The situation is only slightly different for soft X-rays. We see from Fig. 3.4 that when $\lambda = 10$ Å the value of the attenuation length $1/\alpha$ is still very small, of the order of 10^{-5} m.

On the other hand, for wavelengths below 1 Å the absorption coefficient becomes very small, corresponding to weak absorption and to a large attenuation length: the beam can penetrate in the material for at least 1 cm. At even lower wavelengths, the absorption becomes essentially negligible.

In addition to this universal trend as a function of the wavelength, the absorption of X-rays is characterized by other general properties. In particular, the absorption increases with the density of the solid.

Furthermore, it changes with the atomic number and the atomic weight, becoming stronger for materials formed by 'heavy' atoms like lead than for those containing 'light' atoms like beryllium. The causes of this trend must be found in the X-ray absorption mechanisms at the atomic level, discussed later.

Extensive databanks are available through the Internet to evaluate the absorption properties of elemental and composite materials in a wide range of wavelengths. We suggest, in particular, the already mentioned excellent site of the Center for X-ray Optics at the Lawrence Berkeley Laboratory [http://www-cxro.lbl.gov].

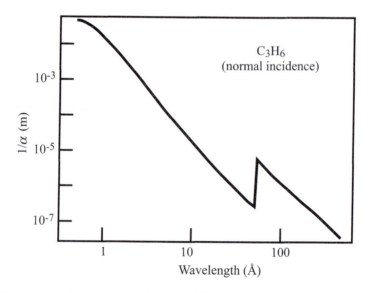

Fig. 3.4 X-ray attenuation length as a function of the wavelength (in a log–log plot) for solid polypropylene. Data from the Center for X-ray Optics of the Lawrence Berkeley Laboratory [http://www-cxro.lbl.gov].

3.1.1.3. Index of refraction

Refraction is a well-known phenomenon for visible light: a beam reaching the interface between two transparent solids (or air and a solid) changes its direction. This is explained by the change in the speed of light as the propagating medium changes—see Fig. 3.5. In vacuum (or, approximately, in air), the speed of light equals c. In a solid, it changes to c/n; the quantity n is the 'index of refraction'.

The relation between wavelength and frequency (eqn 1.1) gives in vacuum $\lambda = c/v$. In a solid, the relation becomes $\lambda = (c/n)/v = c/(nv)$. This implies that either the wavelength changes from vacuum to a solid, or the frequency changes (or both). For visible light, the frequency corresponds to the color. When colored light enters a transparent material such as a colorless glass, its color does not change. Thus, the frequency is the same in vacuum and in a solid, and the wavelength changes from λ to λ/n.

This change has important implications extending beyond refraction phenomena. For example, we found (Section 2.1.1) that the directions of maximum-intensity diffraction by a grating depend on the wavelength. If the grating is immersed in a transparent medium rather than in air, then the wavelength changes and so do the diffraction effects. This conclusion can be generalized: all diffraction and interference phenomena change from vacuum to a transparent medium because of the wavelength change.

The concept of index of refraction can be extended from visible light to other spectral ranges, and specifically to ultraviolet light and to X-rays. However, the n-values for X-rays are very close to unity: the speed of X-rays is almost the same in a solid and in vacuum.

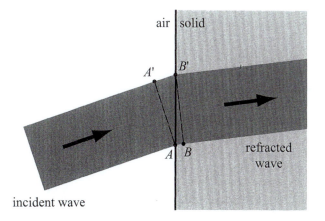

Fig. 3.5 Schematic explanation of the refraction of a visible-light beam at the interface between air and a solid. The speed of light changes from c to c/n. Consider the evolution of the wavefront from AA' to BB': the time needed by the light to travel from A' to B' is the same as the time to travel from A to B. However, the distances $A-B$ and $A'-B'$ are different because the speed of light is different. Thus, the direction of the light beam changes at the air–solid interface.

This point is illustrated by Fig. 3.6, which shows the difference $(1-n)$ as a function of λ for polypropylene—the same material of Fig. 3.4. This plot reveals several interesting facts. The n-value is indeed very close to unity. Furthermore, the $(1-n)$ difference decreases in general as the wavelength decreases, becoming almost negligible for hard X-rays.

Third, a comparison of Figs. 3.4 and 3.6 reveals a clear correlation between the quantities α and n: specifically, 'something happens' at $\lambda = 42$–44 Å both for the absorption coefficient and for the index of refraction. This specific correlation is explained in the next section.

The correlation between α and n is not limited to the case of Figs. 3.4 and 3.6, but constitutes a very general property present for all materials and all wavelengths. It is explained by theoretical optics as the consequence of very fundamental laws such as the principle that a cause must precede in time its effect (see Inset F).

Our present scope, however, does not require a sophisticated theoretical treatment of this point. The important fact to keep in mind is that the two quantities α and n are related to each other, so that a phenomenon affecting one of them must correspond to a related phenomenon for the other, as suggested by the comparison of Figs 3.4 and 3.6.

The same comparison also reveals another important point. Both α and the difference $(1-n)$ become smaller as the wavelength decreases towards hard X-ray values. However, the decrease is *faster* for α than for $(1-n)$. For example, when λ changes from 300 to 1 angstrom, α decreases by nine orders of magnitude (from 10^7 m to 10^{-2} m), whereas $(1-n)$ decreases by seven orders of magnitude (from 0.1 to 10^{-6}).

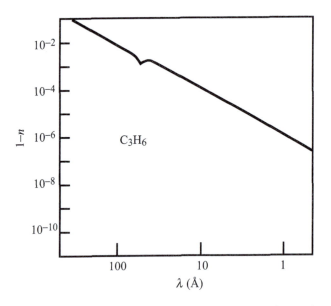

Fig. 3.6 Deviation from unity of the X-ray refractive index of solid polypropylene as a function of the wavelength (in a log–log plot). Note that 'something happens' to n at the same wavelengths for which a marked edge is present in the α-plot of Fig. 3.4. Data from the Center for X-ray Optics of the Lawrence Berkeley Laboratory [http://www-cxro.lbl.gov].

This fact has important consequences for radiological analysis and for X-ray imaging in general. Radiology is based on the capability of X-rays to penetrate into biological materials, i.e. on the low absorption coefficient of such materials. On the other hand, the contrast in radiological images is due to differences in the absorption coefficient between different parts of the object.

Since the absorption is weak, these differences are also weak—and so is the contrast. As a consequence, it is often difficult to obtain good radiological images without large X-ray doses.

Could this problem be solved by exploiting for radiology the refractive index rather than absorption? This is an intriguing idea: since α decreases at short wavelengths faster than $(1-n)$, the differences in n between different parts of the object are typically more marked than those in α. This intuitive idea of using the refractive index is the basis of several innovative radiological techniques, discussed later in this section. Such techniques could, in the forthcoming years, lead to a true revolution in radiology.

Inset F: The complex index of refraction

The quantities α (absorption coefficient) and n (index of refraction) as well as their links can be better understood by using a slightly more advanced theoretical description of electromagnetic waves. Consider (Fig. F-1) a beam of X-rays with only one wavelength λ, corresponding to a collimated and monochromatic plane wave. Suppose that the wave travels along the positive direction of the x-axis.

An electromagnetic wave is a propagating perturbation of the electromagnetic field with electric and magnetic components. The electric field component of the wave of Fig. F-1 is in a direction perpendicular to the x-axis, since all electromagnetic waves are 'transverse' waves. The magnitude E of this field is a function of position x and of time t:

$$E = E_0 \exp\left[2\pi i\left(\frac{x}{\lambda} - vt\right)\right], \tag{F1}$$

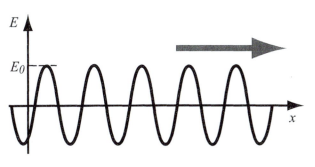

Fig. F-1

where E_0 is the maximum magnitude of the electric field of the wave. Equation F1 corresponds to an oscillating function like that of Fig. F-1, since an exponential function with an imaginary argument is a combination of oscillating sine and cosine functions:

$$\exp(i\theta) = \cos(\theta) + i\sin(\theta) . \tag{F2}$$

Why use imaginary exponentials rather than sine or cosine functions? The answer is that imaginary exponentials are mathematically simpler to handle and more effective.

The propagating nature of the wave is implicit in eqn F1. Consider, in fact, the wave $E(x,t)$ found at a given position x and at a given time t. As the time changes, the wave changes too, according to eqn F1. Consider, in particular, the wave found at the position $(x+\Delta x)$ and at the time $(t+\Delta t)$. This is the same wave found for x and t if the two arguments of the corresponding exponential functions (eqn F1) are equal:

$$2\pi i\left(\frac{x}{\lambda} - vt\right) = 2\pi\left[\frac{x+\Delta x}{\lambda} - v(t+\Delta t)\right],$$

which gives

$$\frac{\Delta x}{\Delta t} = \lambda v . \tag{F3}$$

This result has a simple meaning: the wave found at position x and at time t propagates with speed λv. In vacuum, this must coincide with the speed of light; thus

$$\lambda v = c ,$$

which is equivalent to eqn 1.1.

What happens if the wave propagates in vacuum? The speed changes from c to c/n, the wavelength from λ to λ/n, and eqn F1 becomes:

$$E = E_0 \exp\left[2\pi i\left(\frac{nx}{\lambda} - vt\right)\right]. \tag{F4}$$

Consider now a material that is not completely transparent but partly absorbing, where the wave propagates but its intensity decays progressively. This case can also be described by a function like that of eqn F4. However, the refractive index is no longer a real number n but a complex quantity N, with a real part n and an imaginary part n_i:

$$N = n + i\,n_i. \tag{F5}$$

Equation F3 thus becomes:

$$E = E_0 \exp\left\{2\pi i\left[\frac{(n+n_i)x}{\lambda}\right] - vt\right\} = E_0 \exp\left[2\pi i\left(\frac{nx}{\lambda} - vt\right)\right]\exp\left(\frac{-n_i x}{\lambda}\right).$$

$$\tag{F6}$$

This equation, although seemingly complicated, is really not difficult to understand. On the right-hand side, the first factor,

$$E_0 \exp\left[2\pi i\left(\frac{nx}{\lambda} - vt\right)\right],$$

is equivalent to the right-hand side of eqn F4 and describes the propagation of the wave in the material. The second factor of eqn F6,

$$\exp\left(\frac{-n_i x}{\lambda}\right),$$

is a decreasing function of the distance x. Multiplied by the first factor, it produces a progressive attenuation of the wave, corresponding to absorption by the material.

More specifically, electromagnetism shows that the intensity of the wave is proportional to the square $|E|^2$ of the electric field of the wave. The square of a complex quantity is, by definition, the product of the quantity itself times its complex conjugate, so that eqn F5 gives:

$$|E|^2 = \text{constant} \times \exp\left[2\pi i\left(\frac{nx}{\lambda} - vt\right)\right]\exp\left(\frac{-n_i x}{\lambda}\right)$$

$$\times \exp\left[-2\pi i\left(\frac{nx}{\lambda} - vt\right)\right]\exp\left(\frac{-n_i x}{\lambda}\right) = \text{constant} \times \exp\left(\frac{-2n_i x}{\lambda}\right), \tag{F7}$$

which corresponds to an exponential decrease in the intensity equivalent to eqn 3.1. The comparison of eqns F6 and 3.1 clarifies in fact the meaning of the imaginary part n_i of the complex refractive index N: the quantity $2n_i/\lambda$ equals the absorption coefficient α.

In summary, the complex refractive index N describes both the wave propagation with a different speed with respect to vacuum and its progressive attenuation due to absorption. The (real) refractive index n and the absorption coefficient α are linked to the real and imaginary parts of N.

The analogy with other complex quantities in physics explains the relation between α and n, suggested by the comparison of Figs. 3.4 and 3.6. In order to illustrate this point, consider for example an electric circuit and the corresponding relation between the applied voltage and the current. If the voltage generator is continuous (DC) and the circuit consists of a resistance, then the current is determined by the resistance value R, which is a real quantity.

If instead (Fig. F-2) the voltage generator is alternate (AC) and the circuit contains inductances and/or capacitors, then the resistance must be replaced by the impedance Z, which is a complex quantity with a real part and an imaginary part. The impedance determines two different properties of the AC current: its maximum amplitude and its phase difference with respect to the AC voltage.

These two properties are not independent but linked to each other. This is clearly suggested by Fig. F-2: close to the 'resonance frequency' v_0 of the circuit, 'something special' happens both to the current amplitude and to the phase difference (which goes through zero). This correlation is reminiscent of the link between absorption and refraction suggested by Figs. 3.4 and 3.6.

The similarity between the two cases is not a coincidence but the result of a deep analogy. Both in the case of the complex refractive index and in that of the impedance, one complex quantity is used to describe two correlated phenomena. A complex quantity of this type is known as the 'response function' of the system where the phenomena occur. The complex refractive index is the (optical) response function of the material, and the impedance is the (electronic) response function of the circuit.

Response functions possess some universal properties. Under very general conditions (the principle that an effect cannot occur before its cause and the linearity of the response), the real and imaginary parts of a response function are linked to each other by equations known as 'Kramers–Krönig (KK) relations'.

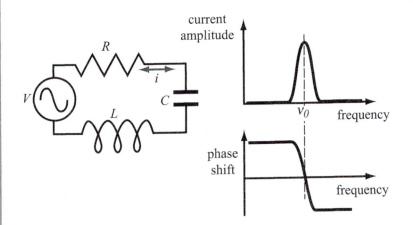

Fig. F-2 An AC circuit with a capacitor (C), a resistance (R) and an inductance (L). The (complex) impedance describes two phenomena: the frequency dependence of the amplitude of the current, and that of the phase shift between current and voltage.

Such relations are beyond the scope of this book: we mention here only one of their implications. A 'special behavior' of the absorption coefficient α as a function of λ (such as a rapid change) must also correspond to a 'special behavior' of n.

This property has an important impact on synchrotron light techniques. Every experimental technique based on absorption has a 'partner' technique based on the refractive index. The KK relations link the data produced by one technique to those of the other, allowing cross checking and/or the derivation of one type of data from the other.

Consider, for example, the visible-light optical effect known as 'Faraday rotation', which is a rotation of the polarization direction caused by the properties of n. The 'partner' technique is 'dichroism': different absorption for differently polarized light. The KK relations link the data from Faraday absorption and from dichroism, so that one set of data can be used to check or derive the other.

In everyday life, the KK relations explain why to fabricate mirrors we coat glass with a strongly absorbing material like a metal. In the case of synchrotron X-rays, the KK relations concern for example the so-called 'anomalous diffraction' (see Section 3.5) and other domains such as ultraviolet absorption and reflection.

3.1.1.4. Microscopic phenomena

The propagation and absorption phenomena of electromagnetic waves in solids are caused by microscopic mechanisms occurring at the atomic level. Such mechanisms are primarily linked to the properties of electrons in the material. Therefore, before discussing specific mechanisms we must present a short and elementary summary of such properties.

3.1.1.5. The electronic structure: from atoms to molecules and solids

Quantum physics provides a coherent description of electronic properties starting from individual atoms. The Z electrons in an isolated atom (Z = atomic number) occupy (Fig. 3.7) a series of discrete energy levels, conventionally labeled as 1s, 2s, 2p, 3s, 3p, 3d, 4s, 4p, 4d, 4f, etc. The integer number in these labels identifies the group of levels or 'shell' to which the level belongs. The conventional names of the shells are: K-shell (1s level), L-shell (2s and 2p levels), M-shell (3s, 3p and 3d levels), N-shell, etc.

The number of electrons occupying each level is limited by the Pauli (exclusion) principle, requiring that no more than two electrons (with opposite 'spins') occupy a given level. It should be noted, however, that p-levels like 2p or 3p are not really single levels: each is a group of three levels, and therefore can accommodate up to $3 \times 2 = 6$ electrons. For similar reasons, each d-level can host up to 10 electrons, and each f-level up to 14 electrons.

The maximum number of electrons, therefore, is different for different shells: 2 for the K-shell, 8 for the L-shell, 18 for the M-shell, etc. Among all natural chemical elements, the one with the largest number of electrons per atom is uranium, with Z = 92 electrons distributed in the K-, L-, M-, N- and O-shells.

The formation of chemical bonds between individual atoms is due to radical changes affecting their highest-energy electrons, which are called 'valence electrons'. There are several types of such changes, corresponding to broad classes of chemical bonds.

In the case of *ionic bonds* (Fig. 3.8) the atomic states of the valence electrons are not substantially changed by the bond formation, but their occupancy is. One or more electrons are transferred from atoms of some elements to atoms of other elements.

The electron-losing atoms become in this way 'positively charged ions' or 'anions', and the electron-receiving atoms become negatively charged ions or 'cations'. The electrostatic attraction between anions and cations creates the chemical bonds.

This attraction, however, does not cause the collapse of anions into cations. This in fact is prevented (Fig. 3.9) by repulsion forces between the atoms that are linked to the Pauli principle. In fact, the collapse of two atoms into one atom would force four electrons into the 1s state, four into the 2s state, 12 into the 2p state, etc., whereas the maximum occupancies according to Pauli are two, two, six, etc. The system reacts against these forbidden occupancies with a strong repulsive force. The interplay between attractive and repulsive forces determines the equilibrium anion–cation distances, that constitute the *chemical bond lengths*.

The same repulsive forces are active in the second broad class of chemical bonds: the *covalent bonds*. However, the attraction between atoms is due in this case to radical changes in the states of the valence electrons, rather than only of their occupancy. Consider for example (Fig. 3.10) the 1s states of the valence electrons of two hydrogen atoms: in the hydrogen molecule (H_2), the two states are combined ('hybridized'). The new state or 'molecular orbital' has a large fraction the electronic charge in the region between the two nuclei. This causes the attraction between the nuclei.

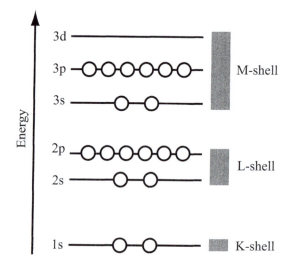

Fig. 3.7 Schematic diagram of the lowest electronic energy levels in an atom (argon) and of the corresponding shells. The maximum number of electrons allowed by the Pauli principle is shown for the 1s, 2s, 2p, 3s and 3p levels, whereas the 3d level in this case is empty.

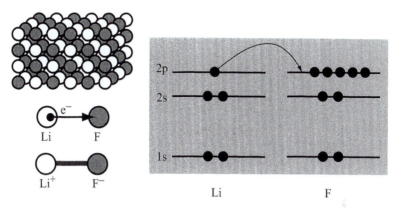

Fig. 3.8 Schematic of the formation of ionic bonds between lithium (Li) and fluorine (F) atoms, leading to the formation (top left) of a LiF crystal. Bottom left: one electron is transferred from a Li atom to a F atom. The Li atom becomes a positively charged ion, and the F atom a negatively charged ion. The electrostatic attraction between these charged ions is the cause of the bond between them. Right: the same electron transfer shown in terms of the energy levels involved.

Besides covalent and ionic bonds, there exist other classes of bonds. *Metallic bonds* (Fig. 3.11) correspond to a strong delocalization of the valence electrons. In hydrogen bonds (particularly important for polymers), the binding action is primarily due to the mediating role of hydrogen atoms. In molecular solids, individual molecules are bound together by weak chemical bonds, primarily of the 'Van der Waals' type.

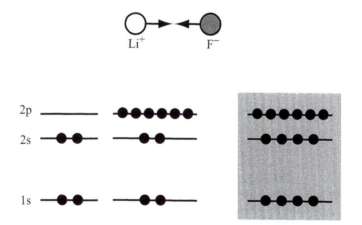

Fig. 3.9 Cause of the repulsion force that makes it impossible for the attractive force between two ions (top) to cause their collapse into one another. As shown in the bottom, after the hypothetical collapse of a Li$^+$ ion into a F$^-$ ion the combined atom would have four electrons in the 1s level and four in the 2s level, in violation of the Pauli principle. The system reacts against this violation with a repulsive force that prevents the collapse.

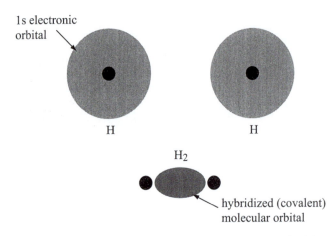

1s electronic orbital

H H

H_2

hybridized (covalent) molecular orbital

Fig. 3.10 Formation of the covalent bond that links two hydrogen atoms (top) in a H_2 molecule. The valence electrons of the two separated atoms occupy 1s atomic orbitals. The hybridization of these orbitals produces (bottom) a molecular orbital that concentrates the electronic charge in the region between the nuclei, energetically favorable for negative charges because of the influence of the positive charges of the nuclei. The result is an attractive force between the nuclei.

The mechanisms of chemical bond formation are quite similar in molecules and in solids. However, the number of atoms and electrons in a solid is much larger than in a molecule. This causes some specific and important properties. In particular, instead of occupying discrete levels the valence electrons in solids occupy 'bands', resulting from the combination of very many individual levels.

The formation of chemical bonds changes of course the energy of the valence electrons. In fact, it is precisely the valence-electron energy decrease that makes the bond formation energetically favorable.

More limited changes occur for the non-valence electrons that occupy lower atomic energy levels. These 'core electrons' are not directly engaged in the formation of chemical bonds. However, their energy levels are indirectly affected by the redistribution of the valence electronic charge caused by the chemical bond formation.

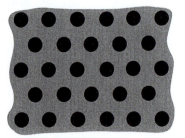

metallic bonds

Fig. 3.11 In a metallic solid, the chemical bonds correspond to a strong delocalization of the charge of the valence electrons.

Consider, for example, an ionic bond between a cation and an anion. After the bond formation, the core electrons of the anion no longer interact with the negative charge of the transferred electron. Thus, they 'see' less negative charge, and this decreases their energy.

Symmetrically, the core electrons of the cation 'see' more negative charge, and this increases their energy. A similar analysis would explain the core-electron energy changes for covalent bonds.

The core-level energy changes due to chemical bond formation are very important for many different synchrotron-base techniques. In fact, core electrons can act as probes of the chemical bond properties, without the complications that affect valence electrons because of their direct involvement in the bond formation.

3.1.1.6. X-ray emission from solids

The existence of discrete electron energy levels explains the emission of X-rays by solids. Imagine a solid bombarded by a beam of electrons. The bombardment can extract core electrons from the atoms in the solid, emptying the corresponding energy levels (Fig. 3.12).

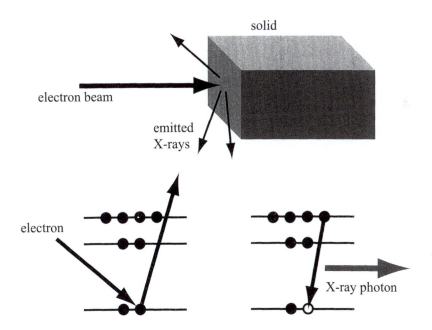

Fig. 3.12 Mechanism of X-ray emission by a solid (top) bombarded by an electron beam. The bombardment (bottom left) can extract core electrons, leaving empty deep energy levels. These deep levels are quickly re-occupied by upper-level electrons. Such electrons lose energy in the form of X-rays photons (bottom right). Because of the large energy difference between the involved electron energy level, the photon energy is typically quite large in the X-ray spectral range.

This is of course a very unstable situation since any transition of an upper-level electron to an empty lower level would decrease the energy of the system. Thus, such transitions do occur very rapidly. The corresponding decrease in energy can lead to the emission of a photon. Because of the large differences between core-electron energy levels, the photon has rather high energy, corresponding to wavelengths in the X-ray region.

This mechanism specifically explains the different spectral distribution (intensity vs λ) of the X-rays emitted by different materials. As shown in Fig. 3.13, a spectral distribution or 'emission spectrum' mainly consists of very strong and narrow peaks, called 'emission lines'. Each line occurs at a characteristic wavelength, corresponding to a photon energy $h\nu$ equal to the jump between two electron energy levels.

Note that the electron energy levels and their differences are specific properties of the atoms of each chemical element. Therefore, the presence of a given line in an X-ray emission spectrum reveals the presence of the corresponding element in the emitting solid. This fact is exploited by several important techniques for chemical analysis.

3.1.1.7. X-ray absorption

The absorption mechanism of X-rays by a solid is closely related to that of X-ray emission, and once again to the electronic levels of its atoms. Figure 3.14 shows a characteristic 'absorption spectrum'—the absorption coefficient of a solid as a function of the photon energy [or of the wavelength, linked to the frequency ν and therefore to the photon energy $h\nu$ by eqn 1.1 ($\lambda = c/\nu$)]. The spectrum exhibits a characteristic 'sawtooth' lineshape.

Figure 3.15 explains the origin of this lineshape. The absorption of an X-ray photon occurs if its energy can be transferred to a core electron. Consider, for example, a core electron in the K-shell (1s level) of one of the atoms in the solid. By absorbing the photon, the electron can leave the atom that thus becomes ionized.

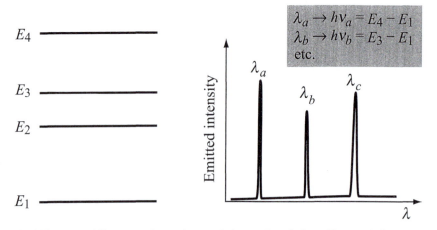

Fig. 3.13 The spectral lines occurring at characteristic wavelengths in an X-ray emission spectrum (right) reflect the distances between the energy levels (left) of the emitting atoms.

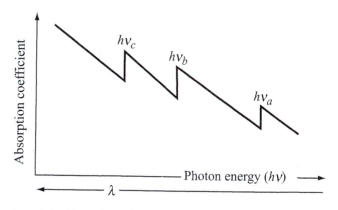

Fig. 3.14 Portion of the X-ray absorption coefficient of a given material as a function of the photon energy (or of the wavelength), exhibiting the characteristic sawtooth lineshape.

The photon energy, however, must be larger than the energy difference between the 1s level and the minimum energy required for ionization. This difference is called the 'ionization threshold' for 1s electrons (K-shell). Other shells are related to other ionization thresholds.

Consider now X-rays with photon energy close to a given ionization threshold. Below threshold, the photons cannot produce ionization and cannot be absorbed with this mechanism. When the photon energy reaches the threshold, there is a sudden jump in absorption due to the corresponding ionization channel. If the photon energy is increased further, the absorption begins to slowly decrease—and so does the absorption coefficient.

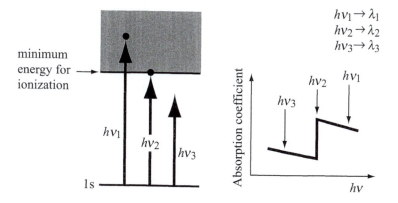

Fig. 3.15 Left: different X-ray photon energies compared with an ionization threshold of a given atom in a material. The photon energy $h\nu_1$ is more than sufficient to produce ionization and thus the photon is absorbed. The photon energy $h\nu_3$ is not sufficient for absorption by ionization. The photon energy $h\nu_2$ is right at the threshold. Right: the corresponding absorption spectrum in the region of the 1s absorption threshold.

Similar to emission, X-ray absorption is exploited in many different analytical techniques. In fact, absorption thresholds are very different for different chemical elements, and their observation can then be used to detect the corresponding elements.

The photoionization absorption mechanism explains why heavy atoms are more effective in absorbing X-rays than light atoms. Compare for example the elements beryllium and lead. A beryllium atom is very light, i.e. its mass is very small. The atomic mass is almost entirely concentrated in the protons and neutrons of the atom nucleus. A small mass implies a small number of protons and therefore a small number of electrons.

Beryllium is in fact a small-Z atom with only 4 electrons, occupying the 1s and 2s levels. These are 'shallow' levels: their distance in energy from the minimum ionization energy is quite small. The corresponding absorption thresholds occur at large wavelengths beyond 111 Å. At lower wavelengths, the photoionization mechanism is not active and X-ray absorption is limited. Beryllium is in fact an excellent material to fabricate windows that must transmit X-rays.

On the other hand, a lead atoms is very heavy (as we can guess from the weight of a piece of lead). Thus, each atom has a large number of protons and electrons. A lead atom possesses in fact 82 electrons, occupying the K-, L-, M-, N- and O-shells. These shells correspond to many X-ray absorption edges (see Fig. 3.16), up to large photon energies and therefore down to small wavelengths (of the order of 0.14 Å).

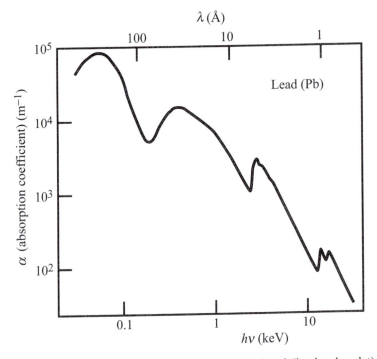

Fig. 3.16 X-ray absorption coefficient as a function of the wavelength (in a log–log plot) for solid lead, a material formed by heavy atoms and therefore a good absorber of X-rays.

This explains why lead is very effective at absorbing X-rays: its absorption thresholds activate the photoionization mechanisms all the way down to small wavelengths. Furthermore, many electrons per atom are active in absorbing X-rays. These arguments justify the well-known fact that lead is an excellent material for X-ray shields.

The photoionization (or photoelectric) mechanism is not the only cause of X-ray absorption. As the wavelength decreases, another mechanism becomes increasingly important: the so-called 'Compton scattering'.

During Compton scattering, an X-ray photon transfers energy to an electron by being deviated, decreasing its photon energy hv and therefore increasing its wavelength. The Compton scattering dominates if λ is shorter than a fraction of one-tenth of an ångström. Even at higher wavelengths it plays a significant role in X-ray absorption.

This role is negative as far as radiology and X-ray imaging in general are concerned. Compton scattering contributes in fact to the diffuse background of scattered photons that deteriorates the image quality and decreases its practical contrast.

3.1.2. Conventional radiology and its limitations

As already mentioned, the contrast in conventional X-ray images is due to absorption—specifically, to absorption differences between different object parts. This requires a compromise between two conflicting factors. To 'see' internal organs the X-rays must penetrate inside the human body without excessive absorption. On the other hand, it is precisely the absorption that makes it possible to see the organs and analyze them.

Weak X-ray absorption means deep penetration but also limited image contrast. Fortunately, the contrast is still sufficient in many cases for radiological analysis. But this is not always true: limited contrast does have a negative impact on important radiology domains such as mammography and angiography.

Consider specifically the case of mammography, whose objective is the early diagnosis of breast cancer in women. This requires detecting radiological details on the scale of a few microns, and therefore a good image quality. On the other hand, the image quality is degraded by the diffuse background caused by Compton scattering and by other factors.

In principle, the image quality could be enhanced by increasing the exposure. This remedy, however, causes major problems in the routine screening of breast cancer. The X-ray dose must be limited to avoid unacceptable risks for presumably healthy patients. Thus, the required image quality cannot be achieved by increasing the exposure and therefore the X-ray dose.

Systematic mammographic screening has thus become a controversial issue. The fear of excessive X-ray doses prevents many women from using this otherwise extremely effective technique, which can sharply reduce the number of deaths and radical surgeries related to breast cancer.

Even without using innovative techniques, synchrotron sources reduce the required X-ray dose for mammography. Contrary to a standard source, a synchrotron can provide specific wavelengths that can be selected over a broad spectral range. The wavelength can thus be optimized for imaging specific parts of the body. Furthermore, the collimation of synchrotron light reduces the diffuse background.

These (almost) automatic advantages are quite useful. However, they do not fully exploit the superior imaging potential of synchrotron sources. More innovative techniques make better use of the unique properties of synchrotron light and are revolutionizing radiological imaging.

3.1.3. Digital subtraction radiology

The first of such innovative techniques is digital subtraction radiology, which is also based on the wavelength-tunability of synchrotron sources. This technique is particularly important in the case of angiography, whose objective is to diagnose problems occurring in very small blood vessels, such as the incipient blockage of a coronary artery.

The diagnosis is complicated by several factors: the small size of the vessels, the similarity of all involved tissues which limits the absorption contrast and the rapid motion of the heart system. Without some help, standard radiology cannot provide an effective solution for these problems.

The image contrast can be enhanced by the injection of a 'contrast agent'. This is a liquid containing the heavy element iodine (with 53 electrons per atom and wavelength absorption thresholds down to approximately 0.37 Å—see Fig. 3.17). The contrast agent absorbs X-rays much more effectively than human tissues which are almost entirely made of light elements (hydrogen, oxygen, carbon).

Unfortunately, the injection of the contrast agent is a complicated and somewhat dangerous procedure. Once a suitable density of contrast agent is achieved in the heart region, the blood flow tends to rapidly eliminate it. Therefore, the contrast agent must be quickly injected right in the heart region using a catheter.

This procedure can causes serious problems such as plaque detachment from the artery walls. The corresponding mortality and morbidity rates are not negligible and make coronary angiography unacceptable for systematic mass screening of heart diseases.

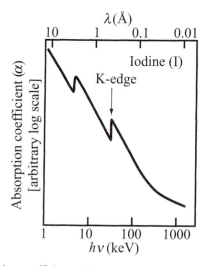

Fig. 3.17 X-ray absorption coefficient of iodine (I) as a function of the wavelength. Note the characteristic K-edge at $\lambda \approx 0.37$ Å, exploited for special radiological techniques.

Synchrotron sources have been used for several years to solve this problem with the digital subtraction technique. This technique is illustrated by Fig. 3.18. Consider the imaging of a blood vessel filled with an iodine contrast agent. As shown in Fig. 3.17, a strong K-shell absorption edge occurs for iodine at $\lambda \approx 0.37$ Å.

Two different radiological images are taken using X-rays with λ-values λ_1 and λ_2 immediately above and immediately below the iodine K-edge. This is not possible with fixed-wavelength conventional sources, but very easy with a tunable synchrotron source. The differences between the two images are due to X-ray absorption caused by iodine. Such differences, therefore, are strong where iodine is present in the first image— notably, the small blood vessels targeted by angiography.

Digital subtraction is the best way to implement this approach in practice. To understand how it works, consider how a black-and-white image is stored in a computer.

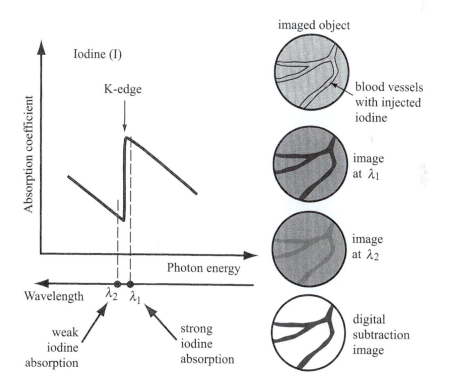

Fig. 3.18 Schematic explanation of digital subtraction angiography. The object (top right: a blood vessel surrounded by other tissues) produces limited contrast by X-ray absorption. The contrast can be enhanced by injecting an iodine-containing contrast agent, exploiting the iodine absorption K-edge (left). With a synchrotron source, the wavelength can be tuned and two images (right, middle) can be taken at two wavelengths, λ_1 and λ_2, immediately above and below the edge. The iodine-related absorption causes a strong difference between the iodine-filled features in the two images. This difference is further enhanced in the digital subtraction image (bottom right).

The image area is divided in small parts (pixels) forming a matrix, and the intensity for each pixel is stored as a number in the image file. The digital subtraction technique consists of creating an image file in which each pixel value is the difference between the pixel values for the two original images.

Compared with the two original radiographs, the digital subtraction picture has sharply reduced background and strongly enhanced iodine-related contrast. The enhanced contrast makes it possible to reduce the concentration of the iodine-containing contrast agent without reducing the image quality. The complicated and risky catheter injection can then be replaced by a much simpler and safer peripheral injection.

This approach was first implemented in the mid-1980s at the Stanford Synchrotron Radiation Laboratory (SSRL), and then developed by several other facilities. Progress has been quite significant, so that synchrotron-based digital subtraction angiography is now almost routinely used for coronary examinations.

3.1.4. Refraction-enhanced imaging

We will now move from absorption-contrast X-ray imaging to other approaches—specifically, those based on the refractive index n rather than on absorption. The simplest one exploits refraction to enhance the edges between different object parts with different n-values. The basic mechanism is explained schematically in Fig. 3.19.

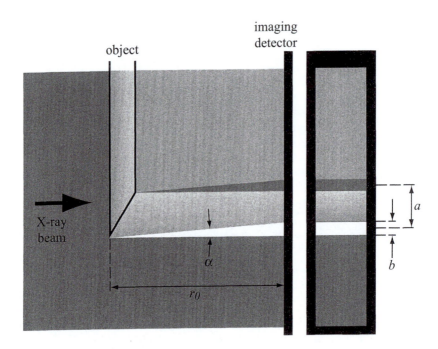

Fig. 3.19 Edge enhancement by refraction: a non-divergent X-ray beam reaches a tapered edge between an object and vacuum. The refraction at the right-hand surface of the tapered edge produces a pair of dark-bright 'fringes' at the detector, thereby enhancing the edge visibility.

Consider an edge between a vacuum (with $n = 1$) and a transparent material (with n slightly different from unity for X-rays). Assume that the edge is not sharp but tapered. Figure 3.19 shows how refraction by the edge produces on the image detector a characteristic sequence of two 'fringes', one dark and the other bright. This sequence enhances the visibility of the edge with respect to standard absorption contrast.

Figure 3.20 shows an example of this approach: a radiographic image with refraction-enhanced edges. The improvement in the edge visibility is clear. Note, in particular, the microscopic size of the imaged features. Figure 3.21 shows another example of the same approach: one of the first refraction-enhanced images of a live specimen.

What are the conditions for obtaining refraction-enhanced images? First of all, the geometry of the X-ray beam plays an important role. The mechanism of Fig. 3.19 is based on a non-diverging beam. With a strongly diverging beam (and/or a beam from a large-size source), the fringes would be blurred and no edge enhancement would be observed.

A quantitative analysis of this point [Y. Hwu *et al.*, *J. Appl. Phys.* **86**, 4613 (1999)] must take into account the parameters labeled a and b in Fig. 3.19, i.e. the distance between the dark and bright fringes and the width of each fringe.

Note that the distance a is determined by the width of the tapered edge, which is of course an intrinsic feature of the imaged object. The parameter a, therefore, cannot be controlled.

On the other hand, the fringe width is given by $b \approx r_0\alpha$, where α is the (small) refraction-induced deflection angle. The width b, therefore, increases with the object-detector distance r_0 and can be controlled by modifying r_0.

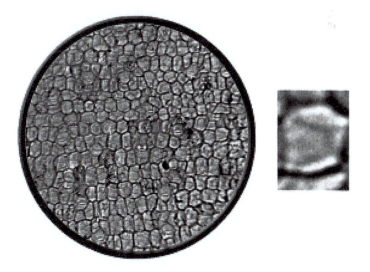

Fig. 3.20 An example of edge enhancement by refraction (from Y. Hwu, Jung Ho Je, Kyu-Ho Lee *et al.*, unpublished results). Left: edge-enhanced microradiology image of a leaf. Right: a magnified detail of the same image, clearly showing the characteristic pairs of dark-bright fringes that enhance the edge visibility.

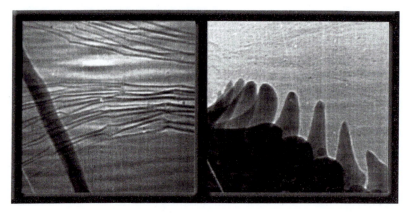

Fig. 3.21 One of the first examples of edge enhancement by refraction for live specimens. The two images reveal microscopic details of a fish (Y. Hwu, Jung Ho Je, Kyu-Ho Lee *et al.*, unpublished results).

The condition for edge enhancement by refraction is to detect the two dark-bright fringes. This requires that $b < a$. When r_0 (and therefore $b \approx r_0\alpha$) become too big, the dark fringe can no longer be seen separated from the bright fringe, and the edge enhancement is weakened or absent.

On the other hand, if r_0 is too small, then the fringe width $b \approx r_0\alpha$ also becomes small, and the detector may not have sufficient spatial resolution to reveal the fringes. In summary, the object–detector distance r_0 must be neither too small nor too large.

The possibility of changing r_0—within the limits of the beamline geometry—adds much flexibility to the technique. In fact, by increasing r_0 we can at a certain point deactivate the refraction-based edge-enhancement mechanism. Thus, such a mechanism can be easily turned on and off, and this facilitates the image analysis.

Similar conclusions are qualitatively valid not only for tapered edges but for any type of non-sharp edge. Furthermore, they are also valid if the edge, rather than separating the object from vacuum, occurs between two different object parts with different refractive indices.

What about the X-ray beam geometry? Figure 3.22 specifically shows the effects of an extended source of size ξ, modeled by two point-sources at a distance ξ from each other. The finite source size 'blurs' the deflection angle α. Calling ρ_0 the source–object distance, the corresponding angular spread is of the order of ξ/ρ_0, increasing with the source size and decreasing as ρ_0 increases. When the angular spread becomes comparable to α, the edge enhancement is no longer visible. This leads to the additional condition $\xi/\rho_0 < \alpha$.

Note that both conditions, $b < a$ and $\xi/\rho_0 < \alpha$, depend on parameters such as a and α that are determined by the microscopic object morphology. Thus, such conditions are not universal but change from object to object and from edge to edge. The general rules are to use a small-size (laterally coherent) X-ray source and a detector with high lateral resolution.

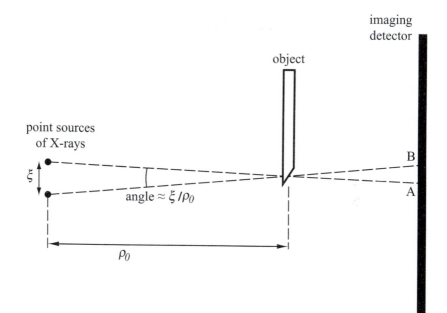

Fig. 3.22 Schematic explanation of how an X-ray source of finite size ξ can jeopardize the refraction-based edge-enhancement mechanism. A finite-size source is modeled here by two point sources at a distance ξ from each other. Each point source produces a pair of fringes for each edge. The two pairs of fringes are centered at the points A and B of the detector, and their angular distance is $\approx \xi/\rho_0$, where ρ_0 is the source–object distance. If this angular spread is small, then the edge-enhancing fringes are still visible. When the angular spread becomes comparable to the α-angle of Fig. 3.19, the edge enhancement is no longer visible.

3.1.5. Phase-contrast imaging

This is another important mechanism—also based on the refractive index—that improves the image quality by edge enhancement. The mechanism is primarily related to the phenomenon of 'Fresnel edge diffraction'. This is a well-known effect in the case of visible-light optics—see Fig. 3.23.

Imagine an opaque object illuminated by a light source with sufficiently high lateral coherence. The object image on the detector includes not only its shadow, but also a series of interference fringes at the object edges. The causes of such fringes are treated in detail in Inset G.

A similar phenomenon is present when the edge is not between a vacuum and an opaque object, but between two transparent objects with different values of the refractive index. Furthermore, the phenomenon also occurs if visible light is replaced by X-rays. In all these cases each edge appears in the image accompanied by a series of bright and dark (diffraction) fringes that enhance its visibility.

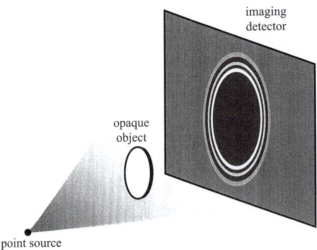

Fig. 3.23 The basic mechanism of phase-contrast edge enhancement: Fresnel edge diffraction. The edge of an opaque object illuminated by a coherent source produces a series of diffraction fringes that enhance its visibility in the image.

This edge-enhancing mechanism is often called 'phase contrast'. An example of phase-contrast imaging is shown in Fig. 3.24. The series of bright and dark fringes is very clear and different from the single pair of dark-bright fringes produced by refraction-based edge enhancement.

What are the conditions for detecting the fringes due to phase contrast? First of all, the X-ray detector must have sufficient lateral resolution to reveal separate fringes. The separation of adjacent fringes (see again Inset G) increases as the square root of the object–detector distance r_0. Thus, a detector with too small a lateral resolution can be improved by increasing r_0—as long as this is allowed by the beamline geometry.

Fig. 3.24 Example of edge enhancement by phase contrast in the radiological image of an optic fiber. Note that each of the two edges of the optic fiber produces a series of fringes, whereas in the refraction-based mechanism only one pair of fringes per edge would be observed.

The second condition concerns the wavelength bandwidth $\Delta\lambda$ emitted by the X-ray source, which must be sufficiently narrow. The reason is that the position of each fringe changes with the wavelength. If the wavelength is 'blurred' by $\Delta\lambda$, then so are the individual fringes. With excessive blurring, the individual fringes become undetectable.

In practice, this condition is quite weak and almost automatically satisfied. One can in fact show (Inset G) that the fringes are still visible if

$$\Delta\lambda/\lambda \ll 2 . \tag{3.2}$$

This constitutes a rather weak condition of longitudinal coherence. Such a condition is automatically met by all synchrotron sources even without using a monochromator to filter the wavelengths. For example, eqn 1.18 shows that it is met by the unfiltered emission of bending-magnet sources.

The weakness of the condition on $\Delta\lambda$ is an important advantage. In fact, one can avoid using a monochromator to reduce $\Delta\lambda$, thus simplifying the beamline, reducing its cost and avoiding monochromator-caused intensity losses. Phase-contrast edge enhancement is indeed routinely observed with 'white' (unmonochromatized) synchrotron X-rays.

The third and final condition for phase-contrast edge enhancement concerns the lateral coherence of the X-ray source—or, specifically, the source size. Imagine, instead of a point-like source, an X-ray emitter of finite size. This source once again can be simulated (as in Fig. 3.22) by a pair of point sources separated by a distance ξ. The two point sources produce two different series of fringes on the imaging detector, shifted with respect to each other. The superposition of the two fringe series can make it impossible to detect individual fringes, and eliminate the phase-contrast edge enhancement.

On the other hand, if the distance ξ is small, then the shift between the two fringe series is also small, and even with the superposition one can still observe individual fringes. It can be shown (Inset G) that ξ is sufficiently small if

$$\xi \ll \rho_0 \sqrt{\frac{\lambda}{r_0}} , \tag{3.3}$$

where ρ_0 is again the source–object distance. This is once again a rather weak condition: using reasonable values for the distances, eqn 3.3 is satisfied by all recent synchrotron sources and by many sources of the previous generations.

In a realistic situation, edge enhancement can be caused both by phase contrast and by the refraction mechanism. Such mechanisms are in fact two aspects of a general phenomenon that can be treated with a unified approach. Our separate analysis for each mechanism does offers some advantages. For example, it easily identifies critical parameters to increase their effectiveness and flexibility.

Specifically, an increase in the object–detector distance r_0 enhances the visibility of the fringes produced by both mechanisms, since they become more separated. We have seen, however, that an increase in r_0 can make the refraction-caused fringes more difficult to detect, whereas no such problem affects the phase-contrast fringes.

Therefore, by changing r_0 we can control the interplay between the two mechanisms, enhancing one or the other as required for specific applications. This interplay is visible in Fig. 3.25: as r_0 changes, we see at first the characteristic pair of refraction-related dark-bright fringes, then an intermediate pattern and finally the series of diffraction-related fringes.

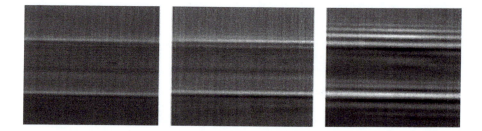

Fig. 3.25 Interplay between the two edge-enhancement mechanisms in the radiological images of an optic fiber. Going from left to right, the detector–object distance r_o is progressively increased making the diffraction-based (phase-contrast) mechanism more prominent with respect to refraction-based edge enhancement—which prevails in the left-hand image. Experimental results derived from: Y. Hwu, H. H. Hsieh, M. J. Lu, W. L. Tsai, H. M. Lin, W. C. Goh, B. Lai, J. H. Je, C. K. Kim, D. Y. Noh, H. S. Youn, G. Tromba and G. Margaritondo, *J. Appl. Phys.* **86**, 4613 (1999).

Note that the object morphology does affect the interplay between the two mechanisms. Specifically, phase-contrast enhancement is present even for very sharp edges, whereas refraction-based enhancement requires an edge of finite width.

By enhancing one mechanism with respect to the other, we change the information content of the radiological image. A phase-contrast image contains more information than a refraction-enhanced image.

Roughly speaking, this information is comparable to a hologram that could be reconstructed with suitable data processing. On the other hand, refraction-enhanced images carry less information but are easier to interpret and therefore more immediately useful for radiological analysis.

Inset G: Image enhancement by Fresnel edge diffraction

The edge between two different parts of an object with different values of the refractive index n can be enhanced in an X-ray image by Fresnel edge diffraction. To describe this mechanism, we begin with the simplest case: an edge between a vacuum and a perfectly opaque object. The analyzed geometry is shown in Fig. G-1.

We can evaluate the detected intensity at each point of the detector by analyzing the different rays that travel from the plane of the object and reach that point. This approach is based on the so-called 'Huyghens–Fresnel principle', widely used in physical optics. The superposition of rays at the detector can produce constructive or destructive interference, thus resulting in bright and dark fringes.

The rules for constructive or destructive interference of two rays are the same as for diffraction gratings (see Fig. 2.4). Two rays tend to cancel each other if their path difference is $\lambda/2$ or an even multiple of $\lambda/2$. Constructive interference occurs if the path difference is λ or an integer multiple of λ.

As the first step, imagine that the opaque object in Fig. G-1 is removed. The combination of all rays would produce uniform intensity at the detector, with no shadow and no fringes. Now, after putting the opaque object back in its place, consider the combination of idealized rays at the detector point z.

The rays blocked by the detector obviously make no contribution to the detected intensity. The rays above the ray 'b' in Fig. G-1 would produce by themselves an illumination equivalent to that of a semi-infinite plane. Thus, they approximately correspond to a homogeneous illumination of the corresponding portion of the detector. As to the rays between the 'a' and 'b' rays, one can imagine combining them in pairs, as shown in Fig. G-2.

Each pair is formed by two rays leaving the object plane at a distance $z/2$ from each other—see, for example, the pairs 'a + c' or 'd + e'. The path difference between two rays in a pair is approximately:

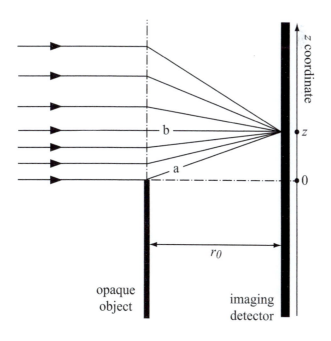

Fig. G-1

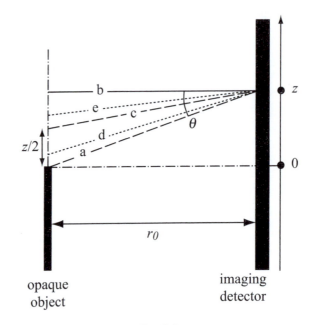

Fig. G-2

path difference $\approx (z/2) \sin\theta$.

In the limit case of small values of z and therefore small values of θ, we can say that $\sin\theta \approx z/r_0$, and

path difference $\approx z^2/(2r_0)$.

Note that this path difference is the same, to a first approximation, for all pairs of rays. As a consequence, the general condition for destructive interference, corresponding to the first dark fringe (path difference = $\lambda/2$), is

$$\frac{\lambda}{2} = \frac{z^2}{2r_0} ,$$

which gives

$$z \approx \sqrt{r_0 \lambda} . \tag{G1}$$

This dark fringe is preceded by a bright fringe, which we can imagine roughly half-way between the origin and the first dark fringe:

$$z \approx 0.5 \times \sqrt{r_0 \lambda} \,. \tag{G2}$$

Other bright and dark fringes follow the first two, forming the overall fringe series.

This analysis is approximate, but its results are not far from reality. Figure G-3 shows the actual intensity plot at the detector as a function of z. The first bright and dark fringes occur at

$$z \approx 0.85 \times \sqrt{r_0 \lambda} \quad \text{and} \quad z \approx 1.3 \times \sqrt{r_0 \lambda} \,,$$

not too far from the rough estimates of eqns G1 and G2. The corresponding first bright–dark fringes distance is:

$$\approx 0.45 \times \sqrt{r_0 \lambda} \,,$$

very close to the value $0.5 \times \sqrt{r_0 \lambda}$ derived from eqns G1 and G2.

What are the conditions for observing the series of diffraction fringes and therefore the edge enhancement? First of all, the detector lateral resolution must be sufficient to separate adjacent fringes.

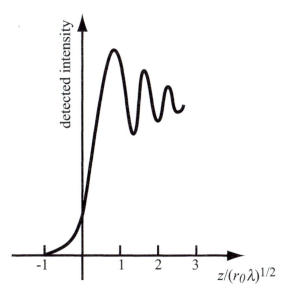

Fig. G-3

The distance, $\approx 0.45 \times \sqrt{r_0\lambda}$, between the first adjacent bright and dark fringes can be increased by increasing the object–detector distance r_0. Thus, a detector with insufficient limited lateral resolution can be compensated by placing it far away from the object—as long as this is not in conflict with the beamline geometry.

The second condition concerns the wavelength, or more precisely the longitudinal coherence of the source. Suppose that the source emits not only one individual wavelength λ, but a band of width $\Delta\lambda$. As a consequence, the diffraction fringes are blurred and may no longer be visible. The blurring Δz can be calculated starting from eqn G1:

$$\Delta z \approx \left(\frac{\partial z}{\partial\lambda}\right)\Delta\lambda = \left(\frac{\partial\sqrt{r_0\lambda}}{\partial\lambda}\right)\Delta\lambda = \frac{1}{2}\sqrt{\frac{r_0}{\lambda}}\Delta\lambda\,.$$

Roughly speaking, the fringes are still visible if the blurring Δz is smaller than the z-value given by eqn G1 for the first dark fringe:

$$\Delta z \ll \sqrt{r_0\lambda}\,, \tag{G3}$$

which leads to

$$\frac{1}{2}\sqrt{\frac{r_0}{\lambda}} \ll \sqrt{r_0\lambda}\,,$$

and therefore to the condition of eqn 3.2.

The third and last condition concerns the finite source size and is related to its lateral coherence. The effects of the source size ξ once again can be modeled (similar to Fig. 3.22) by two point sources at a distance r_0 from the objet and at a distance ξ from each other. The two sources produce on the detector two fringe series shifted with respect to each other by $\Delta z \approx (r_0/\rho_0)\xi$. By applying the condition of eqn G3, we obtain

$$\frac{r_0}{\rho_0}\xi \ll \sqrt{r_0\lambda}\,,$$

and therefore the result of eqn 3.3.

The above analysis applies to the edge between an opaque object and a vacuum. It can be shown, however, that very similar conclusions are valid for the edge between two transparent regions with different values of the refractive index [G. Margaritondo and G. Tromba, *J. Appl. Phys.* **85**, 3406 (1999)].

Specifically, for X-ray wavelengths and very weak absorption the first dark and bright fringes occur symmetrically with respect to the origin, which corresponds to the edge projection on the detector. Their distance is close to $\sqrt{r_0\lambda}$, which is larger that the first dark–bright fringe separation for the opaque object, $0.45 \times \sqrt{r_0\lambda}$. Therefore, the conditions derived above for the opaque object are still generally valid—and in fact slightly weakened.

3.1.4. Other imaging approaches

3.1.4.1. Direction filtering

Coherent X-rays are quite recent and new approaches to coherence-based imaging are still being invented and tested. We discuss here a few additional cases, based on selecting the direction of propagation of X-rays. This 'direction filtering' requires once again a source with good geometrical characteristics, i.e. small size and angular collimation.

Similar conditions apply to all imaging techniques based on the refractive index. Could a conventional X-ray source be used? Such a source is not small and collimated, but large and divergent. These features can be corrected to some extent by using a screen with a pinhole. The price, however, is wasting the X-rays that are blocked by the screen.

A synchrotron source is instead automatically small and collimated, and therefore ideal for refractive-index imaging. Although the use of conventional sources with pinholes is possible, the corresponding intensity is orders of magnitude below synchrotron sources.

A well-defined direction of the X-ray beam is important not only for the beam–object interaction, but also for image detection. Direction filtering before detection can in fact enhance the effectiveness of refractive-index imaging by separating different contrast mechanisms. Figure 3.26 shows a simple example.

Suppose that in a refraction-enhanced imaging experiment a 'direction selector' is placed between the object and the detector. The 'direction selector' is a device that eliminates all the X-rays not traveling in a specific direction, thus preventing them from reaching the detector.

Imagine that the direction selector allows the passage only of X-rays that travel in the same direction as before they interact with the object—see Fig. 3.26b. This eliminates the X-rays that are edge-refracted. The result is an image with absorption as the only contrast mechanism.

If, on the contrary, the direction selector allows the passage (Fig. 3.26c) only of X-rays that are refracted by a given edge, then the background of non-refracted X-rays is eliminated from the image and the edge visibility is enhanced. Furthermore, different edge morphologies produce different refraction angles, and the direction selector could selectively enhance only edges of a given shape.

This analysis clearly shows how a direction selector can separate the imaging effects of absorption from those of the refractive index. But how can a direction selector be fabricated for X-rays? The answer is provided by the Bragg diffraction mechanism of Fig. 2.10.

In fact, a Bragg-diffracting crystal eliminates all X-rays of a given wavelength except those traveling in directions compatible with the Bragg law (eqn 2.3). For each family of crystal planes, a specific direction is thus filtered for each wavelength. Depending on the crystal quality and on the corresponding 'rocking curve', the direction selection can be extremely accurate and within a very narrow angular tolerance.

The example of Fig. 3.26 is very simplified, and more sophisticated applications of direction filtering have been conceived. In particular, a clever and very effective approach called 'Diffraction Enhanced Imaging' or DEI was developed by Chapman *et al.* [D. Chapman, W. Thomlinson, R. E. Johnston, D. Washburn, E. Pisano, N. Gmür, Z. Zhong, R. Menk, F. Arfelli and D. Sayer, *Phys. Med. Biol.* **42**, 2015 (1997)].

The DEI experimental setup is shown in Fig. 3.27. An early example of DEI results is shown in Fig. 3.28.

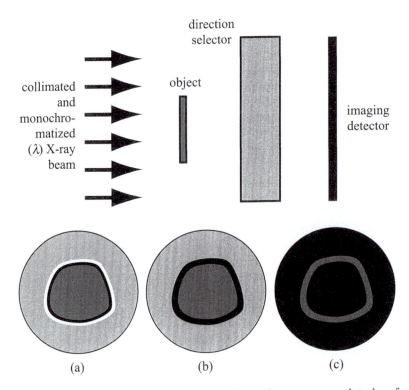

Fig. 3.26 A simple use of direction filtering. The experimental setup corresponds to the refraction-enhanced technique of Fig. 3.19, but the X-rays here are monochromatized (i.e. they have a well-defined wavelength) and there is an additional device before the detector that selects only X-rays traveling in a specific direction. The three images correspond to: (a) refraction-enhanced imaging with no direction filtering; (b) the same image, with direction filtering that eliminates refracted X-rays; (c) again the same image, with direction filtering that allows only refracted X-rays to reach the detector. Images (b) and (c) illustrate the separation of absorption contrast and refraction contrast.

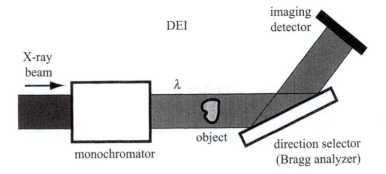

Fig. 3.27 Experimental setup for the DEI technique. This is a practical implementation of a scheme similar to Fig. 3.26, using a Bragg analyzer as direction selector. However, the practical implementation includes data processing based on a rather sophisticated and very powerful mathematical algorithm.

The DEI approach is based on taking images in two directions corresponding to two symmetric points of the rocking curve of the direction selector. A mathematical algorithm is then applied to the corresponding pixels of the two digitalized images. This yields two processed images corresponding to two different image formation mechanisms: absorption and refraction.

The DEI technique is therefore a clever, powerful and sophisticated version of the simplistic approach of Fig. 3.26. The separation of absorption and refractive index effects produces complementary radiographs of superior quality.

Furthermore, even before the DEI mathematical data processing, direction filtering drastically reduced the scattering background in the images. This already produces a substantial enhancement of their quality.

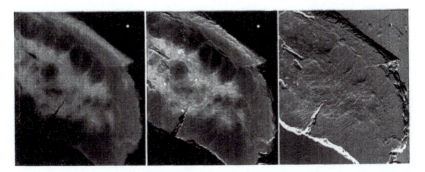

Fig. 3.28 One of the first results of the DEI technique. Left: conventional radiograph. Center: absorption-contrast radiograph. Right: refractive-index radiograph. Details on the results can be found in: D. Chapman, W. Thomlinson, R. E. Johnston, D. Washburn, E. Pisano, N. Gmür, Z. Zhong, R. Menk, F. Arfelli and D. Sayer, *Phys. Med. Biol.* **42**, 2015 (1997). See also: G. Margaritondo, *Phys. World* **11**, 28 (1998).

3.1.4.2. Interferometry

The superior geometrical characteristics of synchrotron X-rays are also exploited by interferometric approaches to imaging. A typical experimental setup for interferometry is shown schematically in Fig. 3.29.

The setup includes three perfect crystals, each one functioning as a phase-coherent beam splitter or as a mirror [for a detailed description see, for example: A. Momose, *Nucl. Instrum. Meth.* **A 352**, 622 (1995); A. Momose *et al.*, *Nat. Med.* **2**, 473 (1996); T. Takeda *et al.*, *Radiology* **214**, 298 (2000)]. The resulting images exhibit interference fringes caused by the phase changes occurring in the object that is placed along one of the two beam paths.

Such fringes contain much valuable information quite close to holography. However, the practical application of this approach requires an extremely accurate alignment of the components of the experimental system. This requirement makes the approach difficult to implement.

3.1.4.3. Medical and materials science applications

The ultimate test of a medical radiology technique must of course be performed on live specimens and specifically on live human patients. Several positive tests of this kind have already confirmed the feasibility and effectiveness of the synchrotron-based techniques described in this chapter.

Synchrotron angiography based on digital subtraction has been already used for hundreds of patients and is well beyond the stage of the first human tests. Similar synchrotron-based mammography tests are underway.

As to refractive-index techniques, extensive tests have been performed on live animals—see, for example, the results illustrated by Fig. 3.21. These tests clearly demonstrated that high contrast and excellent image quality can be achieved in real time on live specimens.

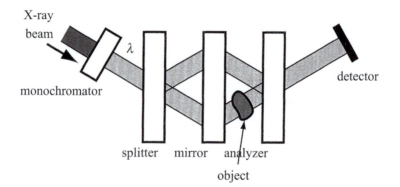

Fig. 3.29 Experimental setup for interferometric X-ray imaging, based on three almost perfect crystals acting as beam splitter, mirror and analyzer.

Furthermore, early tests confirmed that refractive-index radiology can reduce the X-ray dose. In fact, images with comparable quality can easily be obtained with a substantially smaller dose with respect to standard radiology.

In addition, microradiology tests have already shown that refraction-based imaging can be extended to microscopic-scale details, yielding images of small parts of living organs with high accuracy and contrast. This is quite important for specific fields like angiography or mammography, since microanalysis is sometimes required for diagnosis.

The use of novel radiology techniques based on synchrotron X-rays is not limited to the medical domain. Important applications are found in biological research, in particular when the techniques are applied on a microscopic scale.

Equally important applications are found in material science, for example in microdefect analysis for mechanical components. Successful tests were also performed for the study of bubbles and growth products in electrochemical processes—as well as for the investigation of fracture processes in metallic films.

3.2. Spectroscopy

Spectroscopy is a general term identifying experimental techniques that are based on the interaction of the system under investigation with a beam of particles (photons, electrons, neutrons, etc.). Such techniques are implemented by analyzing the same particles after the interaction (or other particles produced by the interaction) to reveal important properties of the system. This broad definition applies to a wide variety of experiments, including many approaches based on synchrotron-emitted photons.

Because of the specific scope of this book, we concentrate our attention on five spectroscopy domains: X-ray absorption in general, circular dichroism, fluorescence, infrared spectroscopy and photoemission. Such techniques in fact play a key role in synchrotron-based analysis, in particular as far as biology and medical research are concerned.

3.2.1. X-ray absorption spectroscopy

The discussion of Section 3.1.1 provides the conceptual background for the use of X-ray absorption for chemical analysis. We have seen, in particular, that each threshold in an absorption spectrum reveals the presence of the corresponding chemical element. This approach, however, extracts only a very limited portion of the chemical information potentially available from X-ray absorption.

Absorption phenomena are in fact more complex than the simple analysis of Section 3.1.1, and therefore carry much more information than this analysis would suggest. This is quite evident from the inspection of a real absorption spectrum like that of Fig. 3.30, referring to the spectral region of the potassium 1s absorption edge.

The overall spectral lineshape corresponds to the absorption threshold. This is not all, however: the edge-like threshold is accompanied by a finer structure. The oscillations occurring at wavelengths ≈ 0.1–0.2 Å lower than the edge are the so-called EXAFS (Extended X-ray Absorption Fine Structure), whose theoretical treatment is discussed later. The fine structure closer to the edge is the NEXAFS (Near-Edge X-ray Absorption Fine Structure), also known as XANES (X-ray Absorption Near-Edge Structure).

What is the origin of the NEXAFS? We must consider once again the basic core-level X-ray absorption mechanism of Figs. 3.14 and 3.15. The first condition for X-ray absorption is that the photon energy can connect the (occupied) core level and some (unoccupied) state above the minimum threshold for photoionization. This necessary condition, however, does not specify the *strength* of the absorption mechanism.

The overall absorption strength is of course influenced by the number of electrons that occupy the initial core level and therefore can participate to the absorption process. In addition, the absorption depends on the density of final (unoccupied) states available for the photon-assisted excitation. Figure 3.31 illustrates a rather extreme case: the band of final states is interrupted by empty gaps that cause drastic decreases in the absorption coefficient. In less extreme cases, there are no gaps but more moderate variations in the density of final states that produce modulations in the absorption spectrum.

Such variations are caused by the chemical bonding process. The chemical bonds involve valence electrons and strongly change their states determining the density of states. This causes the NEXAFS.

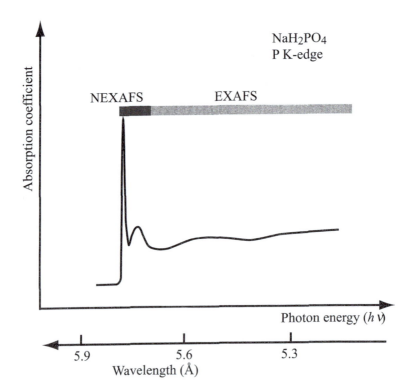

Fig. 3.30 A real X-ray absorption spectrum, showing the edge caused by the excitation of potassium 1s electrons in the compound NaH_2PO_4. Note the fine structure accompanying the edge, including the EXAFS (Extended X-ray Absorption Fine Structure, and the NEXAFS (Near-Edge X-ray Absorption Fine Structure). Experimental data from: Z. Yin, M. Kasrai, G. M. Bancroft, K. H. Tan and X. Feng, *Phys. Rev.* **B 51**, 742 (1995).

Such a conclusion is valid for final-state energies closer than 30-40 eV to threshold. At larger energies, the main effect is no longer the chemical bonding process but the EXAFS mechanism, discussed later. When the final-state energy becomes very large, all modulations fade away.

In addition to the density of final states, other factors affect the strength of the absorption contributing to the NEXAFS. An important one is the so-called 'transition probability'. This is a simple concept: given two electronic energy levels, the absorption of a photon with the right energy (the difference between the two levels) can move an electron from the lower to the upper level. This transition, however, is not guaranteed: depending on the initial and final electronic states, it might even be impossible. In general, each transition has a certain probability of happening that can be calculated using quantum physics. The transition probability determines the relative weight of the corresponding transition in the overall photon absorption process.

Consider for example the 1s electronic state of an isolated hydrogen atom. We could imagine that a photon with the right energy would move the 1s electrons to the 2s level. But this is not true: the 1s → 2s jump is impossible, i.e. the transition probability is zero. On the contrary, the transition probability is non-zero for the jump between the 1s and 3p levels, which can thus contribute to the absorption of photons.

An important point is that the transition probability depends on the polarization of the photon (defined in Section 1.6.). A certain jump may have zero probability for a given polarization and non-zero probability for other polarizations. The dependence of the transition probability on the polarization is determined by the nature of the electronic states involved in the transition. Thus, by varying the polarization and detecting the corresponding changes in photon absorption, we can obtain very valuable information on the electronic states. This approach is particularly powerful when the initial state of the transition corresponds to a core level, since the corresponding initial state properties are quite simple—and this facilitates extracting information on the final state.

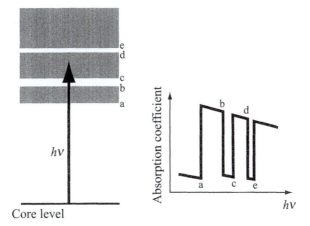

Fig. 3.31 The unoccupied (final) states of the photon-caused 'jumps' can influence the absorption strength and therefore the spectral shape of the absorption coefficient. Here, the final-state band of energies beginning at the threshold (a) is interrupted by gaps (b–c and d–e), which correspond to sharp decreases in the absorption spectrum.

In summary, the NEXAFS of each core absorption threshold primarily reflects the density of unoccupied states; furthermore, the NEXAFS is affected by the transition probability. In turn, the transition probability depends on the nature of the final states. Both the density of final states and their nature are determined by the chemical bond formation process. Hence, X-ray absorption spectroscopy can provide valuable information on the chemical bonds, well beyond the mere detection of chemical elements.

This picture of the absorption process is still somewhat oversimplified and must be completed to take into account several other factors contributing to the NEXAFS. We limit our discussion to three important examples: Cooper minima, excitons and other many-body effects.

A Cooper minimum is a phenomenon affecting the transition probability. For certain core levels, the transition probability becomes zero at a characteristic photon energy, typically a few tens of electronvolts above threshold (see Fig. 3.32).

The reason must be found in quantum physics. Each core electronic state corresponds to an electronic 'wave function'. Certain wave functions become zero (and are said to have a 'node') at a given distance from the corresponding atom. When this happens, the sign of the wave function changes from positive to negative (or viceversa) on going through the 'node'. Quantum physics shows that, as a consequence, the transition probability is zero for a specific photon energy above threshold—and the absorption coefficient exhibits a 'Cooper minimum'.

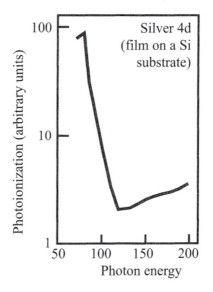

Fig. 3.32 The experimental data for the photoionization cross-section of Ag 4d electrons clearly shows a Cooper minimum. Data from: M. Ardehali, P. H. Mahowald and I. Lindau, *Phys. Rev.* **B 39**, 8107 (1989).

The condition for a Cooper minimum is thus the presence of nodes in the core-level wave function. Only certain atomic states possess nodes: for example, no nodes are found for the 1s, 2s or 3d core states, whereas the 4d and 5f states have one node each (the 3d state has none). Therefore, the detection of a Cooper minimum is a way to check the identification of a threshold as due to a given core state.

The second example of additional NEXAFS factors concerns the so-called 'excitons'. An exciton is a phenomenon due to the interaction between the photon-absorbing electron and the other electrons around it. The interaction is due both to classic electromagnetic effects and to phenomena caused by the Pauli exclusion principle. In spite of their complicated nature, these effects can be described by simple models.

Consider for example the following picture: before absorbing a photon, the system is electrically neutral due to the balance between negative electrons and positive ions. During photoionization, the system loses one negative electron. The system thus becomes positively charged. This positive charge can be visualized as a positive particle, called a 'hole'. During the absorption process, the photon-absorbing electron interacts with the hole that starts to be formed. To a first approximation, the electron and the hole attract each other behaving like the electron and the nucleus of a hydrogen atom.

As shown in Fig. 3.33a, a hydrogen atom has a series of electronic energy levels followed by an ionization threshold and then by a continuum of free-electron energies. All levels below the ionization threshold correspond to an electron bound to the hydrogen nucleus. The continuum above the ionization threshold corresponds instead to the states of a 'free' electron that has left the hydrogen nucleus.

A photon can be absorbed by the electron in a hydrogen atom by stimulating jumps from the lowest 1s ($n = 1$) level to higher energies. The corresponding absorption spectrum (Fig. 3.33b) includes a photoionization threshold preceded by discrete lines. The lines are due to jumps from the 1s level to discrete energy levels. The threshold marks the boundary between discrete levels and the free-electron continuum.

An excitonic electron–hole interaction produces a spectral structure qualitatively similar to that of Fig. 3.33b, with discrete lines below a core-level absorption threshold. This structure contributes to the NEXAFS—see the example of Fig. 3.33c.

Note that excitons can occur only in insulating systems. In metals, many free electrons surround the photon-absorbing atom. The creation of the core 'hole' with its positive charge causes a very fast redistribution of the neighboring free-electron charge which cancels the excitonic effects.

Even in metals, however, the interaction between the photon-absorbing electron and the other electrons contributes to the NEXAFS. The free electrons can absorb small portions of the absorbed photon energy by jumping into slightly excited states.

Thus, the main photon-absorbing electron may not absorb the entire photon energy but a smaller fraction. This influences the NEXAFS in a characteristic way, producing the so-called 'Mahan–Nozières–De Dominicis lineshape'.

Several other factors can influence the NEXAFS, thereby increasing the information content of the absorption spectra but also complicating their analysis. We note, in particular, the influence of atomic-level vibrations through so-called 'phonon broadening'. In general, the level of the data analysis can be calibrated to the type of targeted information.

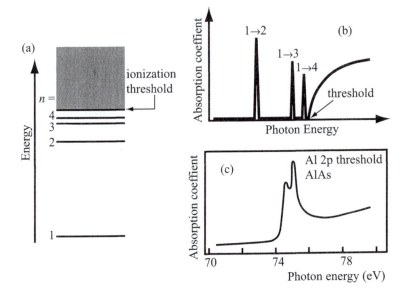

Fig. 3.33 (a) Energy levels and ionization continuum of a hydrogen atom. (b) Corresponding absorption lines for transitions starting from the n=1 (1s) level, followed by the photoionization threshold. (c) Hydrogen-like exciton lines at a core-level threshold [from: M. K. Kelly, D. W. Niles, P. Perfetti, E. Colavita, A. Savoia, G. Margaritondo and M. Henzler, *Phys. Rev.* **B 32**, 5525 (1985)].

The simplest information (the presence of a chemical element) can be extracted from absorption data almost without any processing. More sophisticated information requires advanced processing. As far as biology and the medical sciences are concerned, most experiments require a rather light level of data processing and an elementary understanding of the underlying phenomena—essentially at the level of our present discussion.

3.2.2. Circular dichroism spectroscopy

This is a very important technique for the study of molecules—in particular of biological molecules. It consists of measuring the changes in the absorption spectra when the circular polarization of light is changed.

Elliptical and circular polarization was treated in Section 1.6. (see Fig. 1.16). In essence, an electromagnetic wave is circularly polarized if the direction of its electric field rotates while the strength remains constant, i.e. the wave electric field behaves as a rotating arrow of constant length (Fig. 3.34).

The rotation can occur in two opposite directions: clockwise and counterclockwise, corresponding to 'right-hand' and 'left-hand' circular polarization. We have already mentioned that the transition probability can change with the linear polarization of the photon. This is also true for circular polarization: absorption can be stronger for light with right-hand circular polarization than with left-hand polarization—or viceversa.

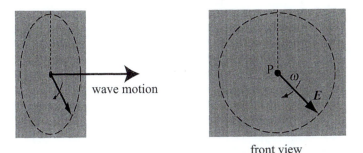

front view

Fig. 3.34 Circularly polarized light: measured at any given point P along the propagation path, the electric field of the wave has constant magnitude and rotates with constant angular speed ω. This means that the tip of the electric field vector moves at constant speed along a circle.

This is the phenomenon called 'circular dichroism'. Such a phenomenon can occur spontaneously due to the intrinsic properties of the absorbing material (see Fig. 3.35). However, it can also be induced by an external magnetic field that causes the so-called 'magnetic dichroism'.

In molecular solids, circular dichroism is related to a property of the component molecules known as 'chirality'. It is also related to a phenomenon called 'ellipticity': the conversion of a linearly polarized wave into an elliptically polarized wave.

The link between dichroism and ellipticity can be understood with a simple analysis of circular polarization. Imagine (Fig. 3.36) two circularly polarized waves with electric fields of equal magnitude. By following the evolution of the vectorial sum of the two electric-field vectors, we realize that the combination wave is linearly polarized.

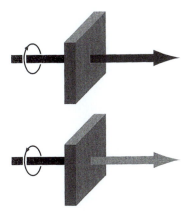

Fig. 3.35 Circular dichroism: the absorption by the crystal is different for left-handed and right-handed circularly polarized light.

Imagine now that the amplitude of one of the two circularly polarized waves is reduced with respect to the other by circular dichroism. Figure 3.36 shows that the combination wave is no longer linearly polarized but elliptically polarized, i.e. its electric field vector tip moves along an ellipse. Thus, circular dichroism transforms linear polarization into elliptical polarization.

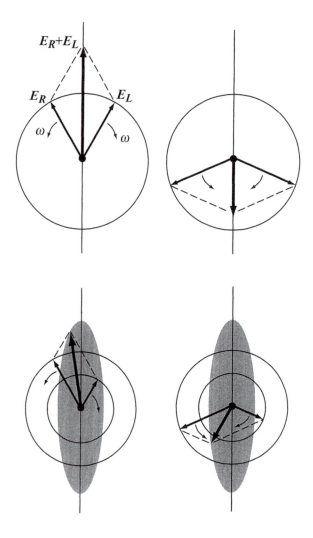

Fig. 3.36 Top: the sum of two circularly polarized waves (left-handed and right-handed) gives linear polarization. Bottom: circular dichroism attenuates the right-handed circularly polarized wave more than the left-handed component giving two arrows of different length. The result is an elliptically polarized wave.

The elliptical character of the polarization is characterized by a parameter called 'ellipticity' (e). This parameter is calculated by first taking the ratio of the minor axis and major axis of the polarization ellipse, and then the inverse tangent of this ratio. Circular dichroism data can be presented in two different ways.

One way is to directly use the difference between the right-hand and left-hand absorption coefficients. The alternate way (see Fig. 3.37) is to use the ellipticity. Either the difference in the absorption coefficient or the ellipticity are plotted as a function of the wavelength (or the photon energy), obtaining a 'circular dichroism spectrum'.

Why is circular dichroism important? The answer is that the chirality of a molecule is directly related to its structure. By analyzing a circular dichroism spectrum, we can obtain valuable information on the molecular structure—complementary to the output of other important techniques such as crystallography and EXAFS (see later).

This point is particularly important for biological macromolecules such as proteins and DNA. Such molecules can in fact exhibit different three-dimensional structures corresponding to different arrangements of molecular elements. Each structure corresponds to a different chirality and produces a distinct circular dichroism spectrum.

For example, different types of secondary structures (helices, sheets, turns and coils) of proteins and other systems can be fingerprinted with the corresponding circular dichroism spectra. Figure 3.38 shows a nice example of this approach.

The differences between right-hand and left-hand absorption produced by chirality are typically very small. Nevertheless, circular dichroism experiments are not extremely difficult.

Even rather small amounts of molecular materials (100 µg or less) can be analyzed rapidly. As a result, circular dichroism is widely used to explore diverse phenomena such as secondary structure in proteins, conformational changes, environmental effects, protein folding and denaturation, as well as dynamics.

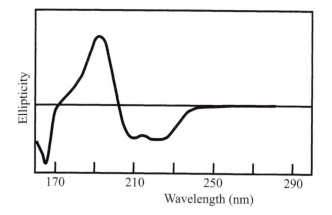

Fig. 3.37 An example of a circular dichroism spectrum: ellipticity as a function of the wavelength for horse myoglobin in aqueous solution Data from B. A. Wallace, *Nature Struct. Biol.* **7**, 708 (2000) and *J. Synchrotron Rad.* **7**, 289 (2000).

Circular dichroism experiment faces a crucial technical problem: the production of circularly polarized light in the wavelength regions of interest. Ultraviolet wavelengths from 1600 to 3000 Å are typically used in the case of proteins and DNA. Strong spectral absorption features occur in fact in this region for the peptide backbone and side chains in proteins as well as for the purine and pyrimidine bases in DNA.

However, the lower end (1600 Å) of the spectral range is determined by the technical limitations of conventional photon sources rather than by scientific arguments. Sources like xenon arc lamps with circular polarizers cannot yield a high intensity at lower wavelengths. As discussed in Section 1.6., synchrotron facilities equipped with special devices such as elliptical wigglers produce very intense circularly polarized waves in a much broader spectral range.

The facilities for circular dichroism based on special synchrotron sources are scarce and in very high demand. Fortunately, the number of facilities is increasing, so that the current limited access problems should be alleviated in forthcoming years.

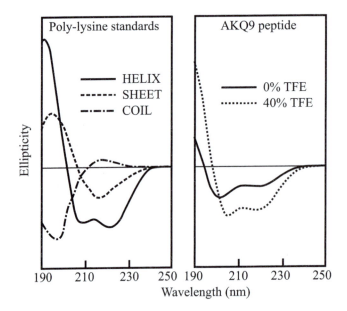

Fig. 3.38 An example of structural analysis using circular dichroism. Left: different circular dichroism spectra for three basic secondary structures of a polypeptide chain: helix, sheet, coil. These standard spectra were theroretically calculated for poly-lysine [N. Greenfield and G. D. Fasman, *Biochemistry* **8** (10), 4108 (1996)]. Right: circular dichroism spectra for the polyglutamine-containing AKQ9 peptide; a theoretical fitting of the solid curve based on the standard spectra indicates a composition with 41% helix and 59% random coil. Adding TFE(trifluoroethanol, a known helix inducer) changes the fit (dotted line) to 71% helix and 29% coil. The results were derived from: E. L. Altschuler, N. V. Hud, J. A. Mazrimas and B. Rupp, *J. Peptide Res.* **50**, 73 (1997).

3.2.3. Fluorescence spectroscopy

The term 'fluorescence' is in most cases identified with the specific X-ray emission mechanism illustrated by Fig. 3.39. An electron is brought to an excited energy level by the absorption of an X-ray photon, and then decays to a lower level emitting another photon. This is, therefore, a slightly different mechanism with respect to the X-ray emission phenomenon discussed in Section 3.1.1 (see Fig. 3.12), in which the electron is brought to an excited level by another electron rather than by a photon.

X-ray fluorescence can be used for chemical analysis in a way similar to electron-stimulated X-ray emission. The emitted photon energies (and wavelengths) are determined by the energies of the involved core levels—which are a characteristic property of each chemical element. The fluorescence results can thus reveal the presence of specific elements as well as their chemical status.

X-ray fluorescence spectroscopy can be implemented in practice in several different ways. The simplest is to measure the emitted intensity as a function of the wavelengths (or of the photon energy)—obtaining 'emission spectra' like that of Fig. 3.40. For this technique the primary photon beam can be either monochromatized (i.e. with a well-defined wavelength) or unmonochromatized ('white').

If a monochromatized primary beam is used, then the emitted X-ray intensity—either at a given wavelength or over a broad wavelength range—can be measured as a function of the primary-beam wavelength (or photon energy). The corresponding plot is a 'fluorescence excitation spectrum'—see Fig. 3.41. This kind of spectrum yields valuable chemical information complementary to that of emission spectra.

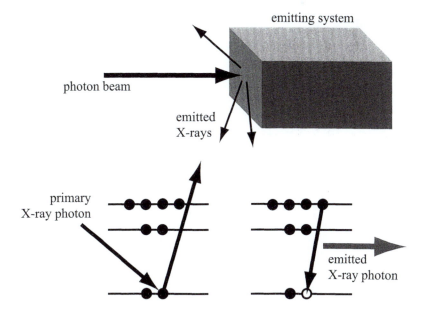

Fig. 3.39 Fluorescence mechanism: the phenomenon is quite similar to that of Fig. 3.12, except for the primary beam that consists of X-ray photons rather than of electrons.

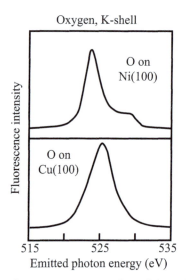

Fig. 3.40 Example of X-ray fluorescence spectra: plots of the emitted intensity vs the photon energy. Results for oxygen 1s electrons (K-shell) in two different oxygen-covered substrates. Data derived from: N. Wassdahl, A. Nilsson, T. Wiell, H. Tillborg, L.–C. Duda, J. H. Guo, N. Mårtensson, J. Nordgren, J. N. Andersen and R. Nyholm, *Phys. Rev. Lett.* **69**, 812 (1992).

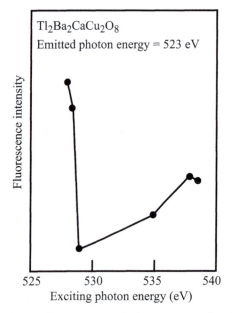

Fig. 3.41 An example of X-ray fluorescence excitation spectrum: plot of the emitted intensity at a fixed photon energy as a function of the primary (exciting) photon energy. The results were obtained on the high-temperature superconductor material $Ti_2Ba_2CaCu_2O_8$. Data derived from: S. M. Butorin, J.-H. Guo, N. Wassdahl, P. Skytt, J. Nordgren, Y. Ma, C. Ström, L.-G. Johansson and M. Qvarford, *Phys. Rev.* **B 51**, 11915 (1995).

The need for a synchrotron source is evident in the case of fluorescence excitation spectra, since they require wavelength scanning and therefore a wavelength-variable source. In general, synchrotrons are better than other sources for all fluorescence spectroscopy modes. They provide primary X-ray beams with superior intensity, thereby increasing the signal level and reducing the chemical detection limits.

This is quite an important point since the fluorescence process is not very efficient. After an electron absorbs a photon and jumps to an excited energy level, it can decay with other mechanisms besides photon emission. For example, it can undergo an 'Auger process'—see Fig. 3.42.

As a consequence, only a small fraction of all excitation processes results in the emission of X-rays. The photon output can be enhanced by an intense primary beam, which is automatically provided by a synchrotron source. Furthermore, the excellent geometrical characteristics of synchrotron X-rays are very important for the microscopy version of fluorescence spectroscopy, discussed later.

Fluorescence spectroscopy can involve either 'deep' core levels or levels close to the ionization threshold. In the first case, the emitted X-rays have large photon energies and short wavelengths. Shallower core levels yield longer-wavelength soft X-rays.

There exists an important difference between these two cases concerning the spectral resolution. This is an important issue since fine chemical analysis must rely on small details in the fluorescence spectra, whose detection requires a monochromator with high spectral resolution.

Even a very high monochromator resolution, however, cannot help if the emitted spectral features are *intrinsically* broad. What can cause this intrinsic broadening? The answer is provided by quantum physics, and specifically by the Heisenberg 'uncertainty' principle.

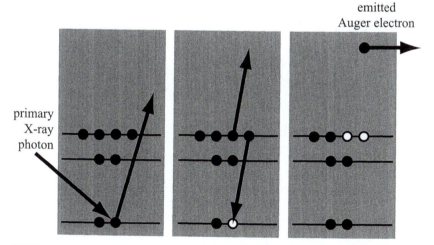

Fig. 3.42 The Auger process provides an alternate mechanism for the de-excitation of an excited electron, in competition with X-ray fluorescence (Fig. 3.39). The core hole is filled by an electron that transfers its energy to another electron thereby enabling it to leave the system as an 'Auger electron'.

The principle states that certain pairs of physical quantities cannot be simultaneously measured with infinite accuracy. This applies, in particular, to energy and time—and specifically to the energy of an electron in a given state and to the total time spent by the electron in the same state.

Suppose that the primary X-ray beam of Fig. 3.39 excites an electron from a 'deep' core level. The low energy of the emptied level makes it very unstable: an electron jumping from a higher-energy level will fill it up very quickly. Therefore, the 'hole' (absence of an electron) in the deep core level has a very short lifetime Δt.

As a consequence, energy measurements cannot exceed an accuracy limit that according to the Heisenberg principle is $\approx h/(\Delta t)$. On the other hand, by measuring the energy of the emitted photon we automatically measure the energy of the 'deep' core level. Therefore, the accuracy limit for the energy level corresponds to an accuracy limit for the photon energy. This is what causes intrinsic spectral broadening: all spectral features must be broader than $\approx h/(\Delta t)$.

When the lifetime Δt increases, the spectral broadening $\approx h/(\Delta t)$ becomes smaller. Such is the case of fluorescence processes involving 'shallow' core levels. Therefore, soft X-ray fluorescence has a smaller intrinsic spectral broadening than low-wavelength fluorescence—and it can more easily be used for fine chemical analysis.

3.2.3.1. Resonant fluorescence

The possibility provided by synchrotrons to change the primary photon energy is specifically exploited to perform 'resonant' fluorescence spectroscopy. What is the difference between resonant and non-resonant fluorescence? The answer can be obtained by considering Fig. 3.43.

In the non-resonant case, the primary photon excites an electron from a core level to an energy well above the minimum excitation threshold. In general, the non-resonant fluorescence intensity does not change very much as the primary photon energy is changed.

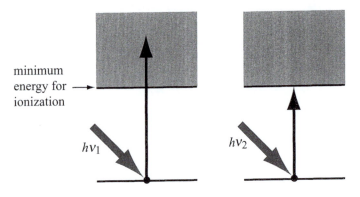

Fig. 3.43 The primary (excitation) steps for non-resonant (left) and resonant fluorescence processes. In the non-resonant case, the photon energy $h\nu_1$ moves the excited electron well above the ionization minimum, whereas resonance (right) occurs when the photon energy $h\nu_2$ is very close to the ionization threshold.

On the contrary, in the case of resonant fluorescence, the primary excitation moves the electron to a state very close in energy to the minimum excitation threshold. This triggers interesting and useful quantum-physics phenomena that can result, for example (Fig. 3.44), in large and rapid changes in the fluorescence intensity vs the primary photon energy. Such changes can be exploited to reveal and study important properties of the system under investigation.

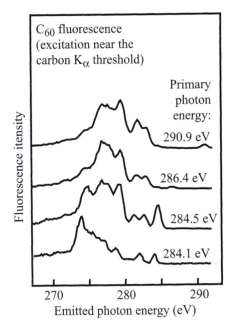

Fig. 3.44 A nice example of resonant fluorescence: the intensity of different spectral features in the emission spectrum changes rapidly as the primary (excitation) photon energy changes. In this case, the emitting material is C_{60} and the exciting photon energies are very close to one of the carbon ionization thresholds. Data derived from: J.-H. Guo, P. Glans, P. Skytt, N. Wassdahl, J. Nordgren, Y. Luo, H. Ågren, Y. Ma, T. Warwick, P. Heimann, E. Rotenberg and J. D. Denlinger, *Phys. Rev.* **B 52**, 10681 (1995).

3.2.3.2. Total reflection fluorescence spectroscopy

This is an interesting variety of fluorescence spectroscopy that is based on the phenomenon of 'total reflection'. Total reflection is linked to the so-called 'X-ray standing waves'.

Total reflection is a well-known effect for visible light. Consider (Fig. 3.45, top left) the interface between a transparent material and a vacuum, and a light beam reaching the interface from the material side. In general, the beam is partly reflected and partly refracted. The refraction theory (see Fig. 3.5 and the corresponding discussion) shows that the refraction angle θ_r is related to the incidence angle θ_i by the simple equation

$$n \sin\theta_i = \sin\theta_r \, , \tag{3.4}$$

where n is the index of refraction of the material. Note that $n > 1$ for visible light, and that $\sin\theta_r \leq 1$. Thus, eqn 3.4 implies that $\sin\theta_i \leq 1/n$. If θ_i is larger than the value corresponding to $1/n$, then no refraction can occur—and the wave is totally reflected.

A similar phenomenon occurs for X-rays, but with a role exchange between the material and vacuum. In fact, for X-rays $n < 1$, therefore total reflection can occur (Fig. 3.45, top right) when the ray reaches the interface from the vacuum side rather than from the material side. Equation 3.4 changes to $\sin\theta_i = n \sin\theta_r$, and the condition for total reflection is $\sin\theta_i > n$.

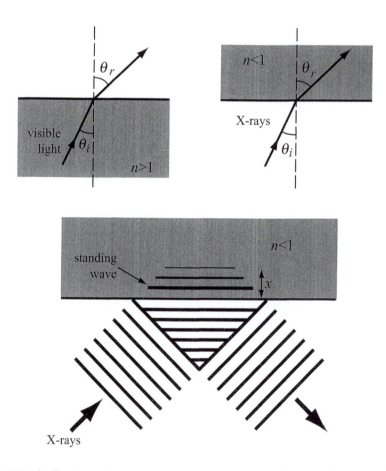

Fig. 3.45 Total reflection and the creation of standing waves. Top left: for visible light ($n>1$), the refraction law (eqn 3.4) predicts a refraction angle θ_r larger than the incidence angle θ_i for a wave moving from the material to vacuum. After θ_r reaches its maximum value ($\pi/2$), an increase in θ_i results in total reflection and no refraction. Top right: a similar conclusion is true for X-rays, but since $n<1$ the wave must travel from vacuum to the material. Bottom: in reality, even with total reflection the X-rays can still penetrate over a short distance $\approx x$ inside the material. The interference between incoming and reflected waves produces a standing wave within the same penetration depth x.

A more detailed analysis of the phenomenon (Fig. 3.45, bottom) shows that even under total reflection the incoming wave can penetrate into the material by a very short distance—but not travel through it. The incoming wave and the totally reflected wave can constructively interfere with each other. The result is a 'standing wave' that does not travel but is stationary and confined near the surface.

This approach enhances the surface sensitivity of the fluorescence probe by concentrating the fluorescence excitation to a thin material slab at the surface. It thus becomes possible to detect small amounts of contaminants at solid surfaces, as required for example by the microelectronics industry. Figure 3.46 shows a nice application of this technique.

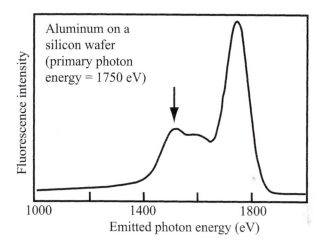

Fig. 3.46 An example of an X-ray fluorescence spectrum taken with high surface sensitivity using the total reflection geometry. Data taken on a silicon wafer intentionally contaminated with aluminum (the vertical arrow identifies the aluminum signal) and derived from: P. Pianetta, K. Baur, S. Brennan, D. Werho and J. Wang, SSRL (Stanford Synchrotron Radiation Laboratory) 1998 Activity Report, p. 7-453. See also: P. Pianetta, K. Baur, A. Singh, S. Brennan, J. Kerner, D. Werho and J. Wang, *Thin. Solid Films* **373**, 222 (2000).

3.2.4. Infrared spectroscopy with synchrotron light

Synchrotron sources are primarily used for the production of ultraviolet light and X-rays. However, from the discussion of Chapter 1 it is clear that they emit electromagnetic waves in many other spectral ranges. For example, a synchrotron source emits intense visible light. This light is not used in practice since other high-quality sources are available that are much less expensive.

On the other hand, synchrotron sources find important applications in the domain of infrared light, including wavelengths from ≈ 0.7 microns to ≈ 100 microns. The reason is evident from Fig. 3.47: when compared with standard infrared sources ('globars' or 'blackbody emitters') a synchrotron is brighter by two to three orders of magnitude.

Many of the infrared applications of synchrotron sources concern the life sciences and are based on absorption spectroscopy. Figure 3.48, for example, shows infrared absorption spectra* of human liver cells taken on a small sample area.

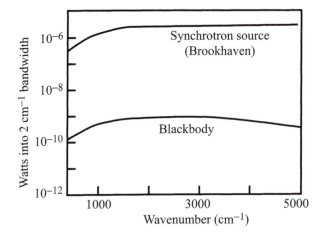

Fig. 3.47 A direct comparison with a conventional (blackbody) source shows the superiority of synchrotrons as infrared sources. The upper curve shows the power delivered into a 10 micron sample by an infrared beamline at the National Synchrotron Light Source in Brookhaven. Data derived from W. Duncan and G. P. Williams, *Applied Optics* **22**, 2914 (1983).

The leading absorption mechanism for infrared light is the excitation of atomic-level vibrations. This mechanism can be understood by considering once again the formation of chemical bonds. Each chemical bond corresponds to an equilibrium distance between the bound atoms. However, the distance between two atoms is not fixed, and can dynamically deviate from its equilibrium value.

The situation (Fig. 3.49) is quite similar to that of a pendulum: the equilibrium position is vertical, and a deviation from equilibrium produces periodic oscillations of frequency v_v. Similarly, the interatomic distances in a molecule or in a solid behave like a set of pendulums or 'harmonic oscillators', vibrating with characteristic frequencies v_v. Each type of chemical bond corresponds to a specific set of frequencies.

(*) Note that the horizontal scale for these spectra is expressed both as 'wavelength' (in micrometers) and as 'wavenumber'—the standard quantity for infrared spectroscopy. The wavenumber is proportional to the photon energy hv and therefore to the photon frequency v. The most frequently used unit is the 'reciprocal centimeter' (cm^{-1}). A photon energy expressed in electronvolts can be converted into a wavenumber expressed in cm^{-1} by using the multiplication factor $\approx 8 \times 10^3$. As to the conversion between a wavenumber and a wavelength, eqn 1.1 [i.e. $v = c/\lambda$] implies that λ is proportional to the reciprocal of the wavenumber. A wavenumber in cm^{-1} is converted into a wavelength in micrometers by taking its reciprocal and multipying it by 10^4. Infrared wavenumbers range approximately from 100 to 14000 cm^{-1}.

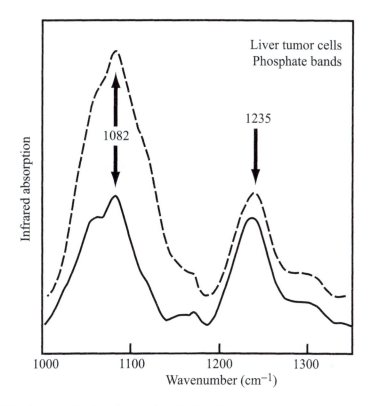

Fig. 3.48 A nice example of synchrotron-based infrared absorption spectroscopy on a microscopic scale at the Advanced Light Source in Berkeley. The study concerns the exposure of human cells to dioxine. The spectral peaks correspond to two stretch vibrations of phosphate bonds in liver cells. A comparison of the results for the reference specimen (solid line) and for a specimen exposed to dioxine reveal large changes in the relative intensity of the symmetric ($1082 \ cm^{-1}$) and asymmetric ($1236 \ cm^{-1}$) stretches. Data derived from: H.-Y. N. Holman, R. Goth-Goldstein, M. C. Martin, M. L. Russell and W. R. McKinney, *Environ. Sci. Technol.* **34**, 2513 (2000).

Quantum physics shows that the energy of an atomic-level oscillation cannot be changed arbitrarily. Like the electrons in atoms, each oscillation has a series of allowed energy levels (see Fig. 3.50). Contrary to the case of atoms, such energy levels are equally spaced from each other—the distance between two adjacent levels being $h\nu_v$.

A molecule or a solid can absorb an infrared photon when one of its atomic oscillations jumps from one energy level to the next higher level. The corresponding photon energy $h\nu$ equals $h\nu_v$.

How can we understand that these photons are indeed infrared photons? Consider that the absorption of infrared light is what warms up our hands when they are close to an incandescent lamp. The warming process cannot be attributed to visible light, since a visible neon lamp does not warm up our hands like an incandescent lamp.

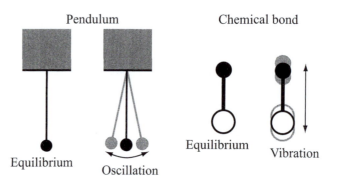

Fig. 3.49 Analogy between the oscillations of a pendulum around its equilibrium point and the vibration of a chemical bond around the equilibrium length.

The photons responsible for the warming mechanism are outside the visible range. Specifically, they must be on the lower-photon-energy side since larger photon energies (ultraviolet light) would break chemical bonds rather than gently exciting their vibrations.

Infrared vibrational spectroscopy is extremely useful for the analysis of molecules and materials because it identifies specific vibrational frequencies and therefore specific chemical bonds. It can reveal, therefore, the chemical elements that are involved in the bonds—and also the detailed bond characteristics.

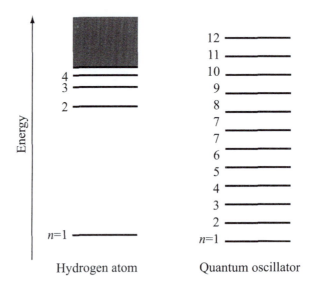

Fig. 3.50 Quantum energy levels for the electron in a hydrogen atom and for an oscillator. Note the main difference: contrary to the atom levels, the oscillator levels are equally spaced.

These capabilities, enhanced by synchrotron sources, are extremely useful in the case of biological specimens. Synchrotron-based infrared experiments are an excellent probe of the microchemical properties of cells and of cellular portions. These techniques can be applied to live cells, and their applications are broad and rapidly increasing.

Biological analysis requires, in most cases, high spatial resolution. Many of the synchrotron infrared experiments are indeed performed in a microscopy mode. The instrumentation to perform infrared microscopy is, in general, much simpler than that for X-rays (see Section 3.3 for a discussion of the problems affecting X-rays).

On the other hand, the larger wavelengths of infrared light with respect to ultraviolet light or X-rays cause some disadvantages. The spatial resolution of conventional infrared microscopy is limited by a fundamental factor: the so-called 'diffraction limit'. This limit will be discussed in Section 4.2.3.

We can anticipate the main point: the maximum resolution achievable by a conventional microscope is fixed by the wavelength of the light. For a conventional infrared microscope, the resolution cannot greatly exceed for example the lower limit of the infrared range, 0.7 microns. However, we will see that this fundamental limitation can be bypassed by the innovative microscopy approach known as 'near-field microscopy' (Section 4.2.3).

3.2.5. Photoemission techniques

Experimental techniques based on the photoelectric effect (or photoemission effect) are among the most widely used in synchrotron-based research. In order to understand why, we must first analyze the basic photoelectric mechanism.

Consider a piece of metal whose high electrical conductivity is explained by the large density of 'free electrons'. What prevents the free electrons from spilling out when they reach one of the outer surfaces of the metal? The answer (Fig. 3.51) is that the electrons find a barrier as they reach the surface. They cannot leave the material unless they have sufficient energy to overcome the barrier.

A closer look at Fig. 3.51 reveals interesting additional points. The free electrons in the metal have different energies rather than the minimum possible energy. This might appear puzzling, since we know that all systems in nature tend to minimize their energy.

The solution of the puzzle is the Pauli exclusion principle, which sets at two electrons (with opposite spins) the maximum occupancy for each electronic energy level. Therefore, the free electrons cannot all collapse into the minimum energy without violating the Pauli principle. The distribution in energy of the free electrons, schematically shown in Fig. 3.51, is a direct consequence of this principle: the electrons fill up the states in pairs, starting from the lowest-energy state and going up to a maximum energy known as the 'Fermi level'.

Even the electrons near the Fermi level do not possess enough energy to overcome the surface barrier and exit the material. To do so, such electrons need an additional energy Φ, known as the 'work function'.

An electron in a deeper energy level needs even more additional energy. If E_b is the distance in energy between the electron and the Fermi level (called the 'binding energy'), then the minimum additional energy required to overcome the barrier is $E_b + \Phi$.

How can this additional energy be supplied? In the photoelectric effect (see again Fig. 3.51) it is provided by the absorption of a photon. If the photon energy is $h\nu$, then a portion $E_b + \Phi$ of this energy is used to overcome the barrier and the rest becomes the kinetic energy K of the electron after it leaves the material:

$$K = h\nu - E_b - \Phi. \tag{3.5}$$

The minimum value of $h\nu$ required to produce photoelectrons corresponds of course to $K = 0$ and to $E_b = 0$ (electrons at the Fermi level). In that case, eqn 3.5 gives $h\nu = \Phi$.

Since Φ is of the order of 5–7 eV, this minimum photon energy is in the spectral range of ultraviolet light (a photon energy of 5–7 eV corresponds to typical ultraviolet wavelengths of 1800–2400 Å). Most photoemission experiments are performed with photons of higher energy and therefore shorter wavelength. Synchrotron sources are excellent for photons of this type.

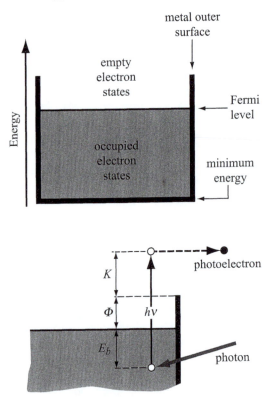

Fig. 3.51 Schematic explanation of the photoelectric effect in a metal. Top: owing to the Pauli principle, the free electrons in the metal occupy a sequence of states at energies up to a level called the 'Fermi level'. Bottom: in order to cross the surface and leave the sample, an electron must overcome a surface barrier Φ (the 'work function'). In this case, the absorption of a photon $h\nu$ by an electron with E_b below the Fermi level brings the electron above the surface barrier, enabling it to leave the sample and become a 'photoelectron' with kinetic energy K.

Equation 3.5 is one of the most famous results of Albert Einstein—marking in 1905 the real birth of quantum physics. It provided indeed a practical way to test Einstein's theoretically-derived hypothesis on the existence of photons. A few years later, photoelectric measurements verified the validity of eqn 3.5 and the notion of photons became universally accepted.

After this fundamental contribution to modern physics, the role of the photoelectric effect became rather insignificant for many decades. In the 1950s it emerged as one of the main probes of the chemical and electronic properties of condensed matter.

The reasons are implicit in eqn 3.5. After the electron leaves the material to become a photoelectron, it becomes possible to capture it and measure its kinetic energy K. From K and using eqn 3.5, we can derive the initial energy of the electron inside the material, corresponding to E_b.

Thus, the energies of electrons in solids or molecules can be retrieved from those of photoelectrons. This capability is exploited in many different ways (see Fig. 3.52).

For example, one can extract photoelectrons from valence electronic states that are directly involved in chemical bonds. The energy distribution of such photoelelectrons (called the 'energy distribution curve' or EDC) reveals the energy distribution of electrons in chemical bonds. This provides extremely valuable and detailed information on the bond formation process.

A similar approach is based on the extraction of photoelectrons from core levels: from their energies one can retrieve the core-level energies. Such energies are characteristic of individual elements. Therefore, specific core-level peaks in an EDC reveal the presence of the corresponding elements. This is the basis of the photoemission technique known as ESCA (Electron Spectroscopy for Chemical Analysis) or XPS (X-ray Photoemission Spectroscopy).

Chemical analysis by photoemission can go way beyond the mere identification of chemical elements: core-level EDC peaks carry a lot of additional information. We saw that electrons in core levels, although not directly involved in chemical bonds, are indirectly affected by the bond formation (see Section 2.1.1.). Their energies are in fact modified by the valence-electron charge redistribution due to the bond formation process. Thus, core-level photoelectron energies can provide valuable information on the valence charge properties and in general on the chemical bonding process.

Figure 3.53 shows an example of this approach. We can see two EDC spectra, each exhibiting a core-level peak. The horizontal scale of these plots corresponds to the core-level binding energy E_b, extracted from the photoelectron kinetic energy K using eqn 3.5 and the known values of the photon energy $h\nu$ and of the work function Φ.

The peaks in the two curves occur at slightly different binding energy values. However, since they fall between 72 and 75 eV, which is close to the binding energy for the Al 2p core level, they can both be attributed to the extraction of photoelectrons from such core levels. This identification is specifically based on tabulated binding energy values. Excellent tabulated data can be found through the Internet, for example at NIST, the National Institute for Science and Technology in the USA [http://srdata.nist.gov/xps/].

The energy difference between the peaks in curves (a) and (b) is due to different chemical bonds involving the aluminum atoms. The top curve (b) was taken on a metallic aluminum film: the valence charge distribution reflects Al–Al metal bonds.

On the contrary, curve (a) was taken on a thinner Al film with much oxidation. The peak position reflects Al–O bonds and the corresponding valence charge distribution. Note that, although much weaker, this same peak is still visible in curve (b) indicating a limited but non-negligible oxidation.

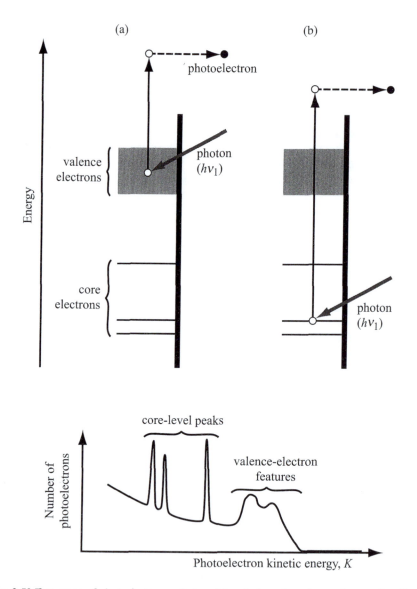

Fig. 3.52 Two cases of photoelectron emission: (a) emission of a valence electron that directly participates in the chemical bonds of the system; (b) emission of a core electron. The two processes produce different features in the EDC spectra (bottom): sharp core-level peaks and much broader valence-electron spectral features

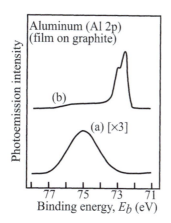

Fig. 3.53 Core-level EDC spectra for two different Al-containing samples. The binding energies identify the peaks as related to Al 2p core electrons. The binding energy is slightly different between curves (a) and (b). Curve (a) corresponds to a thick Al film on graphite whereas curve (b) was taken on a thin (submonolayer) Al film on graphite. The binding energy difference reflects chemical bonding differences for the Al atoms. Data from: G. Faraci, S. La Rosa, A. R. Pennisi, Y. Hwu and G. Margaritondo, *J. Appl. Phys.* **78**, 409 (1995).

The core-level binding-energy changes due to chemical bonds are called 'chemical shifts'. The simplest case is that of chemical shifts due to ionic bonds (Fig. 3.54). The valence-electron charge redistribution corresponds to the transfer of a negative electron between the two bound atoms.

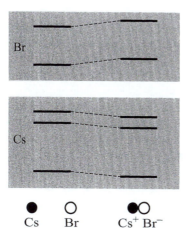

Fig. 3.54 A simple case of core-level chemical shift: the formation of Br–Cs ionic bonds. The core electrons are at lower energies (higher binding energy values) for the negatively charged Br ion than for the neutral Br atom. The opposite is true for the positively charged Cs ion.

Now, the additional electron in the electron-receiving atom 'screens' the positive charge of the nucleus: the core electrons are less attracted by it and their energies are slightly larger than in the unbound atom. The opposite is true for the electron-losing atom. This gives a general idea of how the oxidation state of an atom can be retrieved from photoemission core-level peak positions.

3.2.5.1. Surface sensitivity

We have seen that many decades separated Einstein's theory of the photoelectric effect from its first practical applications to spectroscopy. What caused this long delay?

The answer can be found in surface contamination. Photoemission experiments are highly sensitive to the conditions of the photoelectron-emitting surface: even the slightest amount of contamination strongly affects them.

The surface must be cleaned and then kept clean by not exposing it to gases. Thus, photoemission requires excellent vacuum conditions ('ultrahigh vacuum'). Vacuum technology was not adequate in 1905 and improvements in it were quite slow. Only in the 1950s could the surfaces be adequately cleaned *in situ* and kept uncontaminated. This opened the door to photoemission spectroscopy five decades after Einstein's theory.

Why is photoemission so sensitive to the surface conditions? The point is that an electron, after being excited by absorbing a photon, must travel to the sample surface before becoming a photoelectron. However, excited electrons cannot travel very far from the excitation site before losing energy.

Energy losses are caused by different phenomena: for example, the excited electron can stimulate the creation of 'plasmons', i.e. collective wave motions involving all free electrons. By losing energy, the excited electron may fall below the minimum energy required to pass the surface barrier. In practice, most of the excited electrons do not cross the surface and do not become photoelectrons—and photoelectrons originate from a very thin slab near the sample surface.

The thickness of the electron-emitting slab is called called the 'escape depth'. The escape depth changes with the energy of the photoelectron and therefore (eqn 3.5) with the photon energy. However, the energy dependence is quite similar for different materials and corresponds to the approximately universal curve shown in Fig. 3.55.

We can learn some interesting things from the plot of Fig. 3.55. The escape depth curve exhibits a minimum at 50–200 eV—and the corresponding escape depth is extremely short (of the order of a few ångströms). Therefore, the photoelectrons originate from a near-surface slab no thicker than a very few atomic planes.

Thus, photoemission spectra reflect the electronic properties of the surface rather than those of the bulk material. If the surface is contaminated, then the spectra do not reflect at all the bulk-material properties. That is why very clean surfaces and ultrahigh vacuum are strictly required for photoemission experiments.

Note that surface sensitivity is at the same time a problem and an advantage. Why an advantage? In essence, surface sensitivity enables photoemission to detect surface properties without mixing them with the bulk properties. The surface properties are extremely interesting in fundamental science and also for applications—consider for example corrosion, passivation and catalysis.

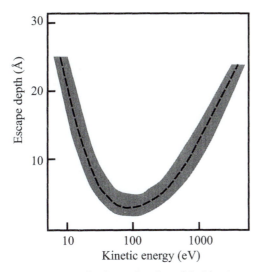

Fig. 3.55 The photoelectron escape depth as a function of the kinetic energy. The dashed line is a typical curve, with its characteristic minimum. The plots for almost all materials fall within the shaded area.

Figure 3.56 shows a nice case of surface-sensitive photoemission core-level spectrum. We see the superposition of two different Pd 3d peaks. Such peaks correspond to the same atomic species—palladium—in two different chemical-bonding situations. The chemical-bonding difference occurs between atoms in the bulk material and atoms near the surface. The difference in energy between the 'surface' peak and the 'bulk' peak is called the 'surface core-level shift'.

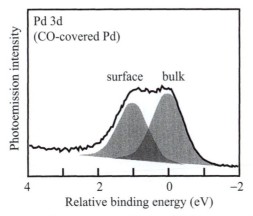

Fig. 3.56 A nice example of surface core-level shift: the Pd 3d spectrum of a palladium sample covered by CO is the superposition of two peaks, one related to the bulk atoms and the other to the atoms near the surface. The shaded areas show a best-fitting estimate of the two components. Data derived from: G. Comelli, M. Sastry, G. Paolucci, K. C. Prince and L. Olivi, *Phys. Rev.* **B43**, 14385 (1991).

3.2.5.2. Why synchrotron light for photoemission experiments?

The most obvious advantage is that intense synchrotron light increases the signal level and therefore the chemical detection capability. But there are other positive aspects.

Synchrotron sources provide tunable photon energy and wavelength, which can be adjusted to meet the requirements of each experiment. Consider once again Fig. 3.55: the photoelectron escape depth—and therefore the surface sensitivity of photoemission experiments—depends on the photoelectron energy. In turn, the photoelectron energy is determined by the photon energy. Thus, a synchrotron source makes it possible to change the degree of surface sensitivity by tuning the photon energy.

Figure 3.57 shows, for example, that spectra taken at different wavelengths exhibit surface and bulk peaks with different relative intensity. With this strategy it becomes easier to identify surface and bulk spectral components, for example.

Taking EDC spectra at different photon energies helps in other ways the difficult process of identifying spectral features. Photoelectrons are created by photon-stimulated excitation of electrons from specific initial states. The probability of the excitation depends on the photon energy in a way that is characteristic of each type of initial state.

In turn, the probability determines the intensity of the photoelectron emission. Therefore, the photon energy dependence of the intensity of a spectral feature can be used to 'fingerprint' the feature as due to a specific initial state.

Figure 3.58 proposes an example of this strategy: the exploitation of a Cooper minimum to 'fingerprint' the corresponding core-level peak. As discussed in Section 3.2.1, Cooper minima in the excitation probability occur for certain core levels at photon energies above the corresponding ionization threshold. Such minima, found in the absorption spectra, also affect the emission of photoelectrons.

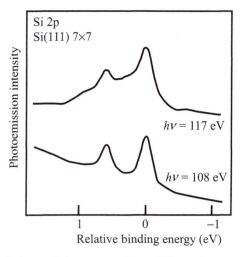

Fig. 3.57 Si 2p core-level photoemission spectra taken at different photon energies exhibit different lineshapes because of the different relative weight of the surface and bulk components. In this case the silicon surface was analyzed with photons of 108 and 117 eV: the latter shortens the escape depth and enhances the surface sensitivity. Data derived from: J. J. Paggel, W. Theis, K. Horn, Ch. Jung, C. Hellwig and H. Petersen, *Phys. Rev.* **B 50**, 18686 (1994).

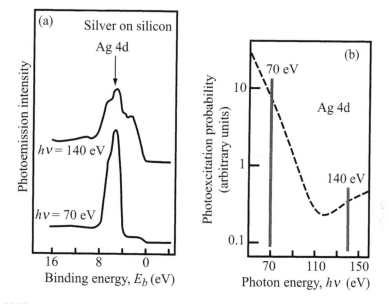

Fig. 3.58 The Cooper minimum technique to identify photoemission spectral features: (a) two EDC spectra taken on a thin Ag film on silicon. As the photon energy is changed from 70 to 140 eV, the relative intensity of the left-hand-side peak decreases dramatically. (b) The effect is due to the Cooper minimum in the excitation probability for Ag 4d electrons: the two shaded lines show that 140 eV is much closer to the minimum than 70 eV. In this way, the spectral contribution of the Ag 4d electrons can be easily identified. Data derived from: J.-J. Yeh, K. A. Bertness, R. Cao, J. Hwang and I. Lindau, *Phys. Rev.* **B 35**, 3024 (1987).

Practically speaking, when the photon energy reaches a Cooper minimum the intensity of the corresponding photoemission peaks decreases. This point is evident in Fig. 3.58: when the photon energy hits the Cooper minimum for initial states of the Ag 4d type, the left-hand-side spectral peak becomes much weaker. This fingerprints the peak as primarily due to the Ag 4d electrons.

More dramatic than Cooper-minimum effects are the so-called 'photoemission resonances', i.e. rapid and strong intensity modifications of certain spectral features as the photon energy changes. The causes are quantum-mechanical, and the resonances happen in general at photon energies near an absorption threshold of the photoemitting atom.

One example of photoemission resonance is seen in Fig. 3.59, showing the rapid changes of one of the features in the valence-electron spectrum of CuO. The resonance occurs at photon energies near the Cu 3p absorption threshold. Specifically note the strong intensity increase of the resonating feature, whereas the resonance does not affect other nearby features.

The usefulness of photoemission resonances is similar to that of Cooper minima: the resonating behavior identifies the nature of the resonating peaks, facilitating the interpretation of the spectra. This is of course particularly helpful for complicated spectra with many overlapping contributions.

3.2.5.3. Non-conventional photoemission modes

In the previous examples the photon energy tunability of synchrotron sources was exploited by simply taking EDC spectra at different photon energies. More sophisticated approaches have been conceived and implemented to take advantage of this property.

In the standard (EDC) mode of photoemission spectroscopy the number of collected photoelectrons is measured as a function of the kinetic energy K, while the photon energy $h\nu$ is kept constant. Imagine, on the contrary, a photoemission experiment in which K is kept constant and $h\nu$ is scanned (see Fig. 3.60). From eqn 3.5, we have:

$$E_b = h\nu - \Phi - K \quad [K = \text{constant}, \; \Phi = \text{constant}];$$

therefore, as $h\nu$ is scanned the binding energy E_b changes too.

Thus, spectra taken in this mode directly reveal the distribution of the electrons in the system as a function of their binding (initial) energy. Since the energy K of the excitation (final) state is kept constant, photoemission data of this type are called 'constant final state' (CFS) spectra.

The main difference between the CFS and EDC modes is that only the initial-state energy changes in the first case, whereas both the initial and final state energies are scanned in the second. This simplifies the extraction of initial-state properties from the CFS curves.

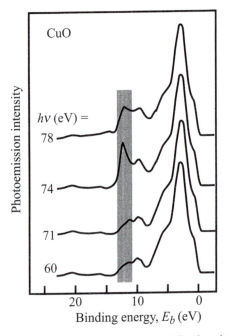

Fig. 3.59 An example of photoemission resonance: the rapidly changing intensity of the spectral feature emphasized by the shaded area. The data were taken on CuO, and the changes occur at photon energies near the Cu 3p absorption threshold. Data derived from: J. Ghijsen, L. H. Tjeng, H. Eskes, G. A. Sawatzky and R. L. Johnson, *Phys. Rev.* **B 42**, 2268 (1990).

The CSF approach is also used to obtain the so-called 'partial-yield spectra'. To understand what these are, we must take into account the following features of the photoelectron emission mechanism. We have seen that most of the electrons excited by photon absorption decay in energy and do not leave the sample. Some electrons are able to leave the sample even after losing part of their energy: as shown in Fig. 3.61, they still retain enough energy to cross the surface barrier.

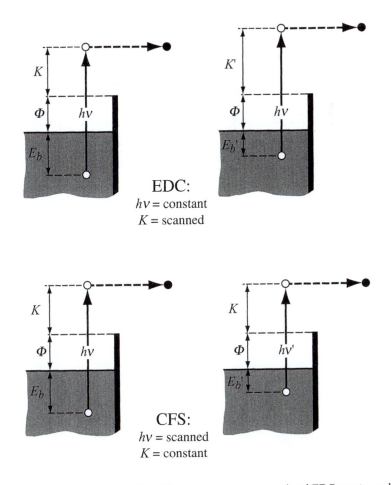

Fig. 3.60 Schematic explanation of the difference between conventional EDC spectra and CFS (constant final state) spectra. In the EDC mode (top), the number of collected photoelectrons is measured while scanning the kinetic energy K and keeping constant the photon energy $h\nu$. In the CFS mode (bottom), $h\nu$ is scanned while keeping constant K. In both cases the binding energy of the initial electron state is scanned (as illustrated by the change from E_b to E_b' between the left-hand and right-hand cases). Thus, both types of spectra reflect the initial-state energy distribution of the electrons. However, the CSF mode is a more direct probe of the initial states since it avoids the simultaneous change of initial and final states occurring in the EDC mode.

A photoemission EDC spectrum thus contains two contributions: (1) the 'primary' photoelectrons, which lose no energy between excitation and emission from the sample; (2) the 'secondary' photoelectrons, emitted after losing part of their energy. The secondary-electron contribution to the EDCs has a characteristic lineshape (Fig. 3.61), with a peak at very low kinetic energies and a long tail extending to high energies. The curve is featureless except for the low-energy peak, so that it does not much complicate the analysis of the primary-electron spectral features in EDC curves.

In the partial-yield mode, $h\nu$ is scanned and K is kept constant at a value coinciding with the low-energy secondary-electron peak. Owing to the strength of this peak, the photoemission signal is almost entirely due to secondary photoelectrons. Thus, a partial-yield curve is essentially a plot of the secondary-photoelectron intensity vs $h\nu$.

What is the information carried by this curve? The key point is that the secondary-photoelectron intensity is *proportional to the photon absorption*. In fact, more absorbed photons mean more excited electrons, and these in turn produce more secondary photoelectrons. Thus, a partial-yield curve essentially *coincides with an X-ray absorption spectrum*, as shown for example by Fig. 3.62.

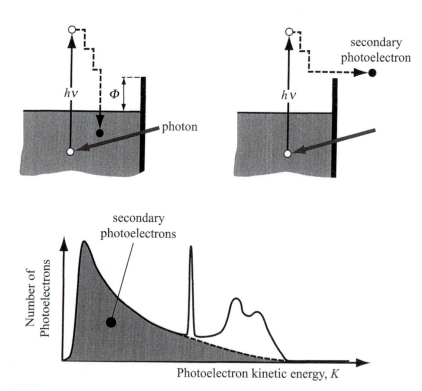

Fig. 3.61 The notion of secondary photoelectrons. Top left: most excited electrons decay in energy and do not leave the sample. Top right: some excited electrons leave the sample after losing part of their energy; these are the secondary photoelectrons. Bottom: the secondary-photoelectron contribution (shaded area) to the EDC spectra is a typical curve with a strong low-energy peak and a long high-energy tail.

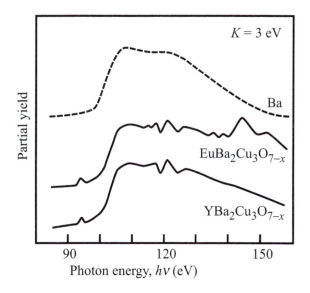

$K = 3$ eV

Ba

$EuBa_2Cu_3O_{7-x}$

$YBa_2Cu_3O_{7-x}$

Partial yield

90   120   150

Photon energy, $h\nu$ (eV)

Fig.3.62 Direct comparison between partial-yield spectra (solid lines) for two barium compounds and the absorption spectrum (dashed line) for barium, at photon energies near the Ba 4d edge. The correlation between partial-yield and absorption is quite clear. Partial yield data from: M. Tang, Y. Chang, M. Onellion, J. Seuntjens, D. C. Larbalestier, G. Margaritondo, N. G. Stoffel and J. M. Tarascon, *Phys. Rev.* **B 37**, 1611 (1988).

Why, however, should we take partial-yield spectra rather than directly measuring absorption spectra? The answer is simple: partial-yield spectra, as all photoemission data, have high surface sensitivity. Therefore, a partial-yield spectrum reveals the absorption coefficient of the surface rather than of the bulk sample. This makes it very useful, notably in surface chemistry and in biology—see, for example, Fig. 3.63.

Finally, we should mention special photoemission techniques that exploit the photon polarization of synchrotron sources. These techniques are based on the fact that the excitation probability depends on the polarization.

More precisely, the dependence of the excitation probability on the polarization is determined by the nature of the initial state. Therefore, by varying the polarization and observing the related changes in the photoemission spectra we can derive very valuable information on the nature of the changing spectral features—see Fig. 3.64.

3.2.5.4. Valence-electron spectroscopy, angle-resolved photoemission

Our discussion of photoemission techniques has thus far focused on core electrons. The reason is simple: core levels are far apart in energy from each other, therefore their identification is easy even in complex spectra—and this facilitates the corresponding chemical analysis. On the contrary, valence electrons occupy rather complicated states resulting from the formation of chemical bonds, and the corresponding spectral features are more difficult to interpret.

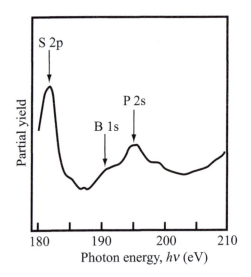

Fig. 3.63 A partial-yield spectrum from a small area of a biological system: different features reveal the presence of physiologic and artificially added elements in a brain anatomical specimen. Data derived from: B. Gilbert, J. Redondo, P.-A. Baudat, G. F. Lorusso, R. Andres, E. G. Van Meir, J.-F. Brunet, M.-F. Hamou, T. Suda. D. Mercanti, M. T. Ciotti, T. C. Droubay, B. P. Tonner, P. Perfetti, G. Margaritondo and G. De Stasio, *J. Phys.* **D31**, 2642 (1998).

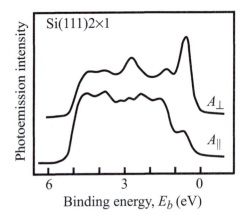

Fig. 3.64 An example of photon polarization effects in photoemission EDC spectra. The results were obtained on a silicon crystal surface. Note the marked differences between the two photon polarizations (marked $A_\perp$ and $A_\parallel$). From these changes one can extract valuable information on the nature of the corresponding electronic states. Data derived from: R. I. G. Uhrberg, G. V. Hansson, U. O. Karlsson, J. M. Nicholls, P. E. S. Persson, S. A. Flodström, R. Engelhardt and E.-E. Koch, *Phys. Rev.* **B 31**, 3795 (1985).

Valence-electron photoemission is nevertheless an important branch of photoemission spectroscopy. The information provided by the corresponding spectra, although difficult to extract, is very valuable because it directly concerns the chemical bonding mechanism. Photoemission experiments on valence electrons are normally performed using ultraviolet photons, with larger wavelengths and lower photon energies than X-rays. The reason is that the intensity of the valence-electron spectral features typically increases as the photon energy decreases. Because of the use of ultraviolet photons, valence-electron photoemission is called UPS (ultraviolet photoemission spectroscopy) rather than XPS or ESCA.

Figure 3.65 shows two typical UPS spectra, taken on a gallium arsenide crystal (top) and on a polycrystalline aluminum film (bottom). The aluminum curve is rather featureless, except for the sharp cutoff that terminates the spectrum.

The origin of this cutoff is the Fermi level, E_F. We have seen (Fig. 3.51) that the Fermi level is the maximum possible energy level occupied by electrons in a solid. In some solids, occupied levels are found all the way up to E_F. In others, there are no allowed states near the Fermi level, so that the highest-energy occupied level is *below* E_F.

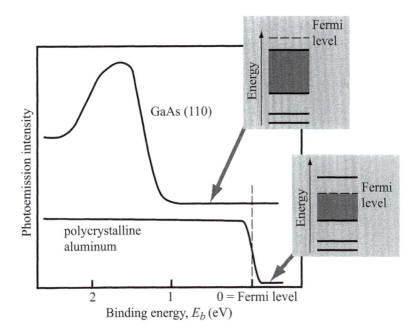

Fig. 3.65 Two examples of UPS spectra for valence electrons. Bottom: spectrum for metallic aluminum. Note the high-energy cutoff, related to the Fermi level. In this case the cutoff is present because the Fermi level of a metal like aluminum falls within a band of allowed states. Top: spectrum for gallium arsenide. The signal does not extend all the way up to the Fermi level: gallium arsenide is in fact a semiconductor and its Fermi level falls within a forbidden gap. Data from: J. Ortega, F. J. García–Vidal, R. Pérez, R. Rincón, F. Flores, C. Coluzza, F. Gozzo, G. Margaritondo, Y. Hwu, L. Lozzi and S. La Rosa, *Phys. Rev.* **B 46**, 10277 (1992).

This difference distinguishes metals from non-metallic samples. A metallic sample is a good conductor: by applying a voltage bias, we can create a current consisting of free electrons. This requires the free electrons to change their state. Such changes, however, imply only small modifications of the electron energy. According to the Pauli principle, they require unoccupied final states with energy close to the occupied initial states.

This condition can be satisfied only if the Fermi level—which separates empty and occupied states—falls in a band of allowed energy levels. If instead the Fermi level is in an energy region without allowed energy levels (a 'forbidden gap'), then the solid is not a conducting metal but an insulator or a semiconductor.

This distinction is quite evident in Fig. 3.65 (see the two insets on the right-hand side of the figure). The top spectrum reveals occupied states all the way up to the Fermi level. On the contrary, in the bottom spectrum there are no occupied states near the Fermi level. This spectrum, in fact, was not taken on a metal but on crystalline gallium arsenide, which is a semiconductor.

Some photoemission spectra exhibit a much richer structure than those of Fig. 3.65. Molecular or molecular-like materials, for example, produce rather complicated spectra with several peaks—see the case illustrated by Fig. 3.66. This is due to the fact that each component molecule brings in the spectra its own characteristic spectral structures consisting of sharp peaks.

Finally, we must mention angle-resolved photoemission. Most photoemission experiments on valence electrons are in fact performed with angular resolution. This means that photoelectrons are detected only along one direction, as shown in Fig. 3.67. The direction can be varied during the experiment: we can thus explore the directional properties of the electronic structure.

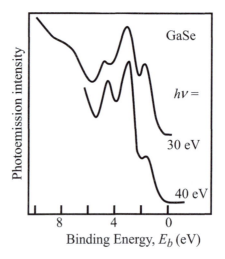

Fig. 3.66 UPS spectra for gallium selenide exhibit a rich structure, reflecting the quasi-molecular character of this material. Data from: G. Margaritondo, J. E. Rowe and S. B. Christman, *Phys. Rev.* **B 15**, 3844 (1977).

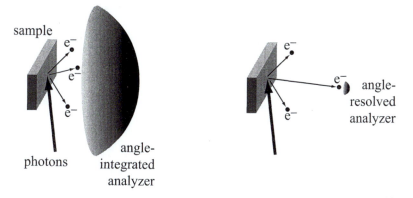

Fig. 3.67 Contrary to angle-integrated photoemission (left), in angle-resolved photoemission (right) only photoelectrons emitted in a selected (and changeable) direction are collected and analyzed.

Imagine, for example, the study of valence electrons in a crystal whose surface has triangular symmetry (see Fig. 3.68). We can expect the electronic structure to have the same symmetry. This is not visible in angle-integrated photoemission experiments that collect photoelectrons in a broad range of directions. On the contrary, the symmetry is strikingly evident in the angle-resolved photoemission results of Fig. 3.68.

The possibility to select the electron direction is one of the strongest aspects of photoemission. Compare, in fact, photoemission spectroscopy and absorption spectroscopy.

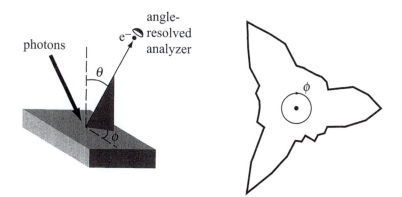

Fig. 3.68 Left: geometry of an angle-resolved photoemission experiment, showing the angles that identify the collected photoelectron direction. Right: a classic example of this approach. The azimuthal plot (as a function of the angle ϕ) shows the intensity of a silicon surface spectral feature. Note the marked three-fold symmetry of this plot, reflecting the similar symmetry of the surface chemical bonds. Data derived from: J. E. Rowe, M. M. Traum and N. V. Smith, *Phys. Rev. Lett.* **33**, 1333 (1974).

The absorption processes probe the initial and final states of the electron excitation. An absorption experiment, however, simultaneously detects the combination of all absorption processes for photons of a given energy, regardless of the directional properties of the corresponding initial and final states.

On the contrary, an angle-resolved photoemission experiment detects the *products* of the absorption processes—the photoelectrons—in specific directions. It can thus explore directional properties not accessible to absorption experiments. Such properties are quite important (see Inset H), and constitute one of the main objectives of advanced photoemission techniques.

Inset H: Angle-resolved photoemission

We have stated that angle-resolved photoemission is important because it probes directional properties that are a fundamental aspect of the electronic structure of solids. In order to see why such properties are so important, we need some additional theoretical background concerning the states of electrons in solids.

Consider an electron in vacuum and imagine that it behaves like a classical point mass. Its state is entirely defined by one physical quantity: the vector velocity v. Other properties can in fact be derived from the vector v; for example, the kinetic energy is given by $mv^2/2$ and the momentum by mv. The vector v is defined by its magnitude (the speed) *and by its direction*. Therefore, the classical-physics state of a free electron already includes an important directional property.

The same conclusion is valid in quantum physics. The free electron is no longer considered as a point-mass, but as an object with wave-like properties. Its quantum wave function is characterized by the wavevector (or k-vector) k, from which other properties can be derived. For example, the momentum is given by $(h/2\pi)k$ and the kinetic energy is

$$\text{kinetic energy} = \frac{h^2 k^2}{8\pi^2 m} . \tag{H1}$$

The wavevector k is characterized by its magnitude k and by its direction, and thus it possesses an important directional property.

When an angle-resolved analyzer captures an electron and measures its (kinetic) energy, it also measures the vector k that defines the free-electron quantum state. In fact, the k direction coincides with that of the captured electron and the k magnitude can be derived from the kinetic energy using eqn H1.

We are not, however, interested in the state of the free electrons captured by the angle-resolved analyzer. Our objective is the state of the electrons in the sample before they are emitted and become photoelectrons. The link between a free-electron state measured by the analyzer and a state in the sample is not trivial—and retrieving one from the other can be a complicated process. Such a retrieval, however, is necessary to extract from the photoemission data information on the electronic structure.

These arguments explain the fundamental difference between angle-resolved and angle-integrated photoemission. Without selecting the direction, photoemission spectroscopy does not yield a complete identification of the state of the photoelectrons—but only measures their energy.

As a consequence, after the retrieval process it cannot provide complete information on the initial quantum state of the electrons in the sample, but only on its energy. If instead the direction is selected, then the state of the photoelectrons can be completely identified. As a consequence, we can retrieve more complete information on the initial state of the electrons in the sample.

This conclusion is illustrated by the example of Fig. H-1, which shows the dependence of the initial-state energy on the k-vector. These $E(k)$ curves—known as 'dispersion curves'—were derived from angle-resolved photoemission data on the semiconducting sample GaSe. Without angular resolution, this information would have been impossible to obtain. The $E(k)$ curves are an extremely valuable tool in understanding the properties of solids: for example, they help to explain optical and electric transport phenomena.

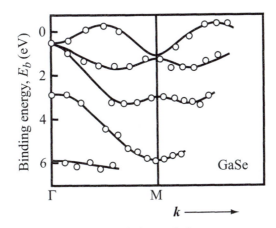

Fig. H-1 Application of angle-resolved photoemission spectroscopy: an early example of experimental mapping of the $E(k)$ ('dispersion') curves. Data for GaSe derived from: P. K. Larsen, G. Margaritondo, M. Schluter, J. E. Rowe and N. V. Smith, *Phys. Lett.* **A 58**, 623 (1976).

In recent years, angle-resolved photoemission has made substantial technical progress, with spectacular improvements in the energy and angular resolution. As a consequence, it can probe very fine electronic properties. In particular, it has made it possible to directly monitor important collective phenomena such as the onset of superconductivity. Figure H-2 shows an example of these novel experimental approaches. Note the shift of the leading edge on the right-hand side of the spectra, which reveals the creation of the 'superconductivity gap' on going from the normal state to the superconducting state .

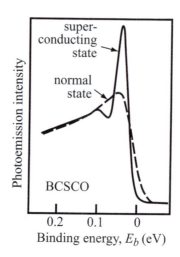

Fig. H-2 An example of angle-resolved photoemission with very high energy resolution: opening of the superconducting gap in a high-temperature superconductor. Data from: Y. Hwu, L. Lozzi, M. Marsi, S. La Rosa, M. Winokur, P. Davis, M. Onellion, H. Berger, F. Gozzo, F. Lévy and G. Margaritondo, *Phys. Rev. Lett.* **67**, 2573 (1991).

3.3. X-ray microscopy and spectromicroscopy

Most of the experimental techniques using synchrotron light were initially developed as macroscopic probes exploring extended sample areas without seeking spatial resolution. Photoemission techniques are a good example; until the late 1980s, they could not analyze areas of size smaller than 0.5 mm.

Applications in the life sciences, in medical research and in many subdomains of chemistry do require spatial resolution. This explains the rapid and accelerating expansion of spatially resolved synchrotron studies.

This expansion involves two related classes of techniques: *microscopy* and *spectromicroscopy*. The scope of a microscopy technique is to produce images with high lateral resolution. In spectromicroscopy, microimaging is coupled to spectroscopic analysis abd is capable of delivering localized chemical and physical information.

Virtually all spectroscopy techniques can be transformed into spectromicroscopy techniques by adding high spatial resolution. The major obstacle is related to photon sources: in most cases, the performances of conventional sources are not adequate for this task.

On the contrary, synchrotron sources are excellent for both microscopy and spectromicroscopy. Their high intensity is helpful in maintaining a reasonably high signal level while enhancing the lateral resolution. The high brightness is also very helpful, specifically while focusing an X-ray beam into a small spot.

3.3.1. Instrumentation and devices: the art of focusing X-rays

Contact imaging is the simplest approach to X-ray microscopy: a beam of X-rays passes through the object, is partially absorbed, and then detected by a photographic plate very close to the object. After developing the film, the image can be viewed with an optical microscope achieving high lateral resolution.

A simple approach of this type, however, does not fully exploit the superior characteristics of synchrotron sources. A more effective approach is to focus the X-ray beam into a small spot—so that synchrotron-based experiments can analyze a small area of the size of 100 Å or less.

Focusing X-rays, however, is not easy. Conventional focusing lenses based on refraction absorb soft X-rays rather than focusing them. Hard X-rays are not strongly absorbed, but refraction focusing is not very effective for them since the corresponding refractive index is very close to unity.

Reflection optics is a good alternative. Spherical, elliptic and toroidal mirrors are indeed widely used as focusing devices. Their major technical problem is that the efficient reflection of X-rays requires a grazing incidence. This corresponds to a wide illuminated mirror area, and achieving high surface quality over a large mirror area is a very difficult task.

We also mentioned that focusing in two perpendicular directions requires toroidal mirrors or (approximately) spherical or elliptical mirrors. An alternate solution is to use a pair of cylindrical mirrors that focus in perpendicular directions. This is the 'Kirkpatrick–Baez' lens mentioned in Section 2.1—see Fig. 3.69.

Other good focusing lenses for soft X-rays are the 'zone plates' and the 'Schwartzschild objectives'. Specifically, the zone plates are similar to the well-known 'Fresnel lenses', widely used for visible light. Figure 3.70 illustrates their working principle.

Fig. 3.69 Top: a cylindrical mirror (working at a grazing incidence) can focus a beam of X-rays only in one direction. Bottom: the combination of two perpendicular cylindrical mirrors produces focusing in two perpendicular directions. This is the device called the 'Kirkpatrick–Baez lens'.

We can see that a zone plate contains a sequence of concentric rings (called 'zones'). In the sequence, an opaque (to X-rays) ring is followed by a transparent ring, followed by an opaque ring, and so on. An incoming X-ray beam can then pass only through the transparent zones. The total X-ray wave after the zone plate is the superposition of the individual waves passing through the transmitting zones. This total wave is strong or weak depending on the path differences between the individual waves.

Consider, in fact, the total wave at point F in Fig. 3.70. The path difference for the two component waves passing through the transmitting zones n and $n+2$ is:

$$\sqrt{f^2 + r_{n+2}^2} - \sqrt{f^2 + r_n^2} \approx \frac{r_{n+2}^2 - r_n^2}{2f}$$

(the approximation is valid if both r_n and r_{n+1} are much smaller than the distance f). If this difference equals one wavelength,

$$\frac{r_{n+2}^2 - r_n^2}{2f} \approx \lambda ,$$

then the two waves combine constructively. This condition is automatically met if

$$r_n \approx \sqrt{n\lambda f} . \tag{3.6}$$

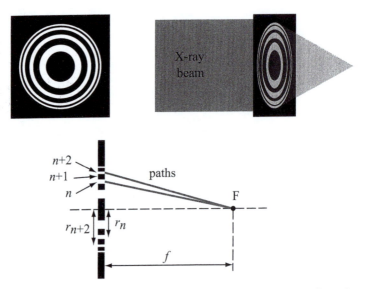

Fig. 3.70 Top left: schematic diagram of a zone plate (with only a very small number of zones shown). Top right: the zone plate acts as a focusing lens for X-rays. Bottom: analysis of the focal conditions (see text).

Note that if the condition of eqn 3.6 is satisfied, then the path difference between any two transmitting zones is always an integer multiple of λ. Thus, eqn 3.6 guarantees that all individual waves combine constructively.

A zone plate fabricated according to eqn 3.6 focuses the beam at the point F whose distance f (the 'focal distance') is implicitly given by the same equation:

$$f \approx \frac{r_1^2}{\lambda}. \tag{3.7}$$

Equation 3.7 shows that the focal distance f changes with the wavelength. If the incoming X-ray beam contains several different wavelengths, then each wavelength is focused at a different point. This could be exploited to separate the wavelengths: zone plates are indeed also used as monochromators.

Zone plate devices do work for visible light. Their fabrication becomes very difficult when the wavelength decreases to the X-ray domain. The problems are primarily due to the narrow width of the zones, which according to eqn 3.6 decreases as n increases. The outermost zone ($n = N$, where N = total number of zones) has the smallest width, given by

$$W_N = r_N - r_{N-2} \approx \sqrt{Nf\lambda} - \sqrt{(N-2)f\lambda} = \sqrt{f\lambda}\left(\sqrt{N} - \sqrt{N-2}\right),$$

which for large values of N can be approximated as

$$W_N \approx \sqrt{\frac{f\lambda}{N}}. \tag{3.8}$$

According to eqn 3.8, the outermost-zone width W_N decreases as λ becomes shorter. When λ reaches the X-ray domain, the zone widths become very small and the zone plate must be fabricated using sophisticated and expensive techniques.

Equation 3.8 also implies that W_N can be increased by decreasing N, the total number of zones in the plate. This, however, is not a good way to keep W_N large, since N cannot be too small. A detailed theory of zone plates shows in fact that the practical width of the focused spot can be improved (i.e. made smaller) by increasing the so-called 'numerical aperture' defined as the ratio between the lens radius and the focal distance. Since the lens radius is approximately equal to r_N—which in turn (eqn 3.6) is proportional to the square root of N—the spot size increases if N decreases. Good focusing requires of course a small spot size, and therefore a reasonably large value of N.

Sophisticated fabrication techniques (such as 'electron beam microlithography') can produce very narrow outer zones with W_N-values of a few hundred ångströms. The corresponding practical size of the focal spot is also a few hundred ångströms, which corresponds to good lateral resolution in microscopy. The corresponding zone plates, however, are small, delicate and difficult to handle.

For the lowest photon energies (largest wavelengths) of the soft X-ray domain, focusing can be achieved using Schwartzschild objectives instead of zone plates. As shown in Fig. 3.71, a device of this kind consists of two spherical mirrors, one convex and the other concave. It can achieve lateral resolution levels comparable with the zone plates.

Schwartzschild objectives are simpler to operate than zone plates. Furthermore, the same Schwartzschild objective focuses both visible light and X-rays, which is an advantage for alignment operations. However, these devices operate only in a limited wavelength (spectral) range.

The narrow spectral range is the consequence of the fact that, as seen in Fig. 3.71, a Schwartzschild objective is based on X-ray reflection at non-grazing angles of incidence. Normally, this would sharply limit the reflectivity and the intensity throughput.

This problem is solved by the special technique of 'multilayer coating'. Figure 3.72 schematically explains how it works. A multilayer coating consists of a periodic series of thin films. When light reaches the coated surface, reflections occur at all interfaces. The total reflected wave is the superposition of the individual reflected waves.

An individual reflection would indeed be weak for X-rays at non-grazing incidence corresponding to a weak total reflected wave. There is, however, an important exception: when the superposition of the individual waves is constructive, the total reflection is strongly enhanced.

As shown in Fig. 3.72, the condition is that the path difference of the two waves reflected by adjacent interfaces equals one wavelength (or an integer multiple of one wavelength):

$$D \sin\theta = \lambda,$$

where D is the multilayer coating period. Note the similarity between this result and the condition for Bragg diffraction by a crystal (Section 2.1.2).

Multilayer coatings are fabricated with alternate depositions of two different materials. Such materials must guarantee the long-term stability of the coating under the heavy thermal load produced by the incident X-ray beam. Only a few kinds of coating materials meet these requirements.

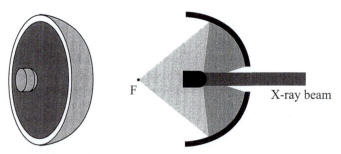

Fig. 3.71 Left: schematic picture of a Schwartzschild objective with its convex and concave spherical mirrors. Right: side view of the same device, showing the reflections that focus the incident X-ray beam.

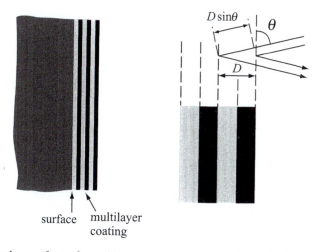

surface   multilayer
                coating

Fig. 3.72 Left: scheme of a surface with multilayer coating to enhance the X-ray reflectivity. Right: derivation of the conditions for reflectivity enhancement.

Those most commonly used materials for soft X-rays contain silicon. This limits the working spectral range to photon energies that are not strongly absorbed by silicon; practically speaking, the upper limit for the photon energy is ≈90 eV. Furthermore, each multilayer coating with a specific period D works for a specific wavelength—or, more precisely, for a narrow band of wavelengths. Thus, each multilayer-coated Schwartzschild objective can in fact only be used in a limited spectral range.

In spite of the spectral limitations, Schwartzshild objectives are widely used for synchrotron-based spectroscopy and spectromicroscopy as well as for other techniques. The performances are quite satisfactory, in particular considering the moderate cost.

Focusing becomes increasingly difficult as the wavelength moves from the soft X-rays domain to hard X-rays. Suitable devices with good performances have nonetheless been developed.

The Fresnel zone plate technology can still be used in the beginning of the hard X-ray region. Furthermore, it can be coupled with Bragg diffraction in the so-called 'Bragg-Fresnel lenses'. These devices are high-quality single crystals—typically, silicon or germanium—with Fresnel zone plates etched on their surface. They thus exploit in parallel Bragg diffraction for monochromatization together with zone-plate focusing.

Figure 3.73 schematically shows one example of Bragg–Fresnel lens. Note that the approach of combining Bragg diffraction and zone-plate focusing can also be implemented using artificial multilayer structures rather than crystals.

Other types of focusing devices for hard X-rays are based on refraction—the same working mechanism of conventional lenses for visible light. The refraction lens technology, however, is very different for X-rays and for visible light, for two reasons. First, the X-ray refractive index of a material is smaller than in vacuum or air ($n = 1$), whereas the opposite is true for visible light. Second, the refractive index value is very similar for a material and for a vacuum (or air).

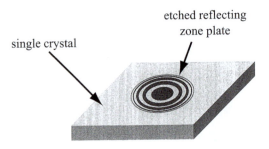

Fig. 3.73 A Bragg–Fresnel lens.

The impact of these differences is made clear by directly comparing the focusing lenses for visible light and for hard X-rays. A typical visible-light lens is made of glass and is double-convex (Fig. 3.74a). The basic lens for X-rays is a double-concave object made of a light element that only weakly absorbs hard X-rays.

As shown in Fig. 3.74b, we could imagine using a piece of boron, carbon or aluminum with concave semi-cylindrical external surfaces. This simple approach, however, encounters serious difficulties. Theory shows indeed that the focal distance is

$$f = \frac{r}{2(n-1)},$$

where r is the radius of the cylindrical surfaces and n is the refractive index. Taking a typical value $(1-n) = 10^{-5}-10^{-6}$, a lens with $r = 0.5$ mm would have a focal distance $f \approx 25-250$ m, which is difficult or impossible to use.

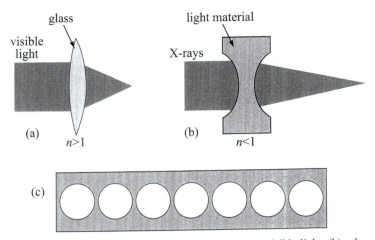

Fig. 3.74 (a) A typical biconvex glass (refractive) lens to focus visible light; (b) scheme of an idealized biconcave refractive lens for X-rays; (c) scheme of a more practical 'compound refractive lens' (CRL) for X-rays.

The solution is provided by devices like that illustrated by Fig. 3.75c: a series of cylindrical holes inside a light chemical element. The focal length is reduced in this case, becoming

$$f = \frac{r}{2N(1-n)},$$

(3.9)

where N is the total number of holes in the array. Devices of this kind are called 'compound refractive lenses' (CRLs). Practical CRLs have tens of holes, thus achieving reasonably short focal lengths. They can focus hard X-rays into spots of the order of a few microns. This approach is also implemented with parabolic rather than cylindrical surfaces, yielding focusing in two perpendicular directions.

In addition to diffraction and refraction, hard X-ray focusing can be obtained by total reflection—with devices somewhat related to optical fibers for visible light. The phenomenon of total reflection for X-rays has already been discussed in Section 3.2.3 (see the top part of Fig. 3.45). We saw that the mechanism is quite similar to total reflection for visible light, except that total refraction occurs when the X-ray beam reaches the material arriving from a vacuum rather than the other way around.

A similar conclusion is valid for the interface between two different materials. Total reflection occurs—beyond a critical angle of incidence—if the X-ray beam travels from the material with a larger n-value to that with a lower n-value.

The n-value depends on the electron density of the material. More specifically, the difference $(1-n)$ is proportional to the electron density. Thus, total refraction occurs when the X-ray beam travels from a 'light' material with small electron density to the interface with a heavier material.

Total refraction is the mechanism that keeps visible or infrared light confined to a waveguide or optical fiber. As shown in Fig. 3.75 (top), a simple optical fiber consists of a glass core surrounded by a glass sheath. Total reflection at the interface between the core and the sheath keeps the light beam within the core and enables it to propagate over a long distance with very limited losses.

A similar technique is used for X-rays, based on thin films. An example of a thin-film waveguide is shown in Fig. 3.75 (bottom). The guiding portion of the structure consists of a polyimide and is surrounded by heavier materials (silicon and silicon dioxide), thus complying with the requirements for the total reflection of X-rays.

Finally, excellent focusing performances can be obtained with tapered glass capillaries, which are also somewhat related to optical fibers. Figure 3.76 shows a schematic explanation of this approach. This technology was the first to achieve (in the mid-1990s) nanometer-scale resolution for hard X-rays.

3.3.2. X-ray microscopy

There exist many different types of X-ray microscopy techniques. Such techniques can be grouped into broad categories based on the corresponding interaction between X-rays and matter: absorption microscopy, reflection microscopy, fluorescence microscopy, etc.

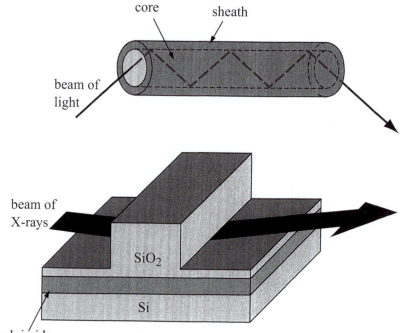

Fig. 3.75 Top: basic scheme of a waveguide (optical fiber) for visible or infrared light. Bottom: a related approach used for X-rays—scheme of a thin-film X-ray waveguide reported in: Y. P. Feng, H. W. Deckman and S. K. Sinha, *Appl. Phys. Lett.* **64**, 930 (1994).

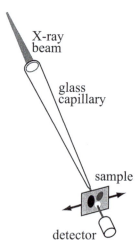

Fig. 3.76 Schematic diagram of an X-ray scanning microscope based on focusing by a tapered glass capillary. For a detailed description see: D. H. Bilderback, S. A. Hoffman and D. J. Thiel, *Science* **263**, 201 (1994).

Furthermore, microscopy techniques can be broadly divided into *scanning* and *non-scanning* approaches. The simplest type of non-scanning technique is shown in Fig. 3.77a: after interacting with the specimen, the X-rays reach a magnifying device (for example, a Fresnel zone-plate lens), and are then detected by a laterally resolving device such as a CCD.

Figure 3.77b shows a more sophisticated non-scanning approach: the X-ray beam is first focused into a small specimen area, thus increasing its local intensity, and then magnified. This is a simple and versatile technique capable of producing high-resolution images within a short data-taking time.

The general scheme of a scanning X-ray microscope is shown in Fig. 3.78. A focusing device concentrates the X-ray beam into a small portion of the specimen. A scanning mechanical device changes the position of the focused spot on the sample surface by moving the sample holder. The detector measures the X-ray intensity for each point of the sample. Two-dimensional (x–y) images can be computer generated by displaying the intensity data in a x–y array.

Both the scanning and the non-scanning approaches are widely used for materials science and for chemical, biological and medical applications. Figure 3.79 shows an example of the applications in materials science.

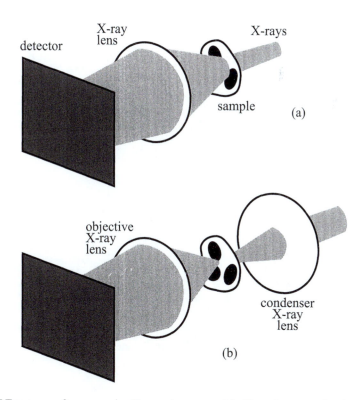

Fig. 3.77 Two types of non-scanning X-ray microscopes. The X-ray lenses can be, for example, Fresnel zone plates.

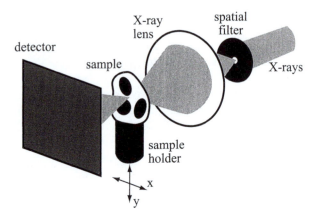

Fig. 3.78 Scheme of a scanning X-ray microscope. The X-ray lens concentrates the X-ray beam onto a small spot on the sample. The sample holder can change the position of the beam on the sample surface, and specifically scan it along the x and y directions to produce two-dimensional images.

Each microscopy technique evolves into a type of spectromicroscopy by achieving high spatial resolution. One good example is the evolution from X-ray absorption spectroscopy to absorption spectromicroscopy. Figure 3.80 illustrates a nice result obtained with this latter technique. Other types of spectromicroscopy are discussed in the forthcoming sections.

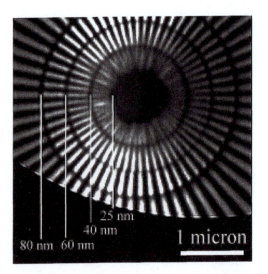

Fig. 3.79 A nice example of transmission X-ray microscopy applied to materials science: the image of a sector disc taken with the Aarhus X-ray microscope, equipped with a zone plate. The width of the spokes is indicated. For details see: R. Medenwaldt and E. Uggerhøj, *Rev. Sci. Instrum.* **69**, 2974 (1998).

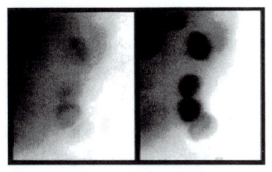

Fig.3.80 Soft X-ray images of dried Chlamydomonas cells taken at photon energies below (left, 533 eV) and at the oxygen absorption resonance (right, 539 eV). Image size: 10 μm × 10 μm. Note the marked change in contrast related to the oxygen distribution in the specimen. Data from: A. D. Stead, T. W. Ford, A. M. Page, J. T. Brown and W. Meyer-Ilse, *Advanced Light Source Compendium of Users' Abstracts and Technical Reports* 1993–96 (Lawrence Berkeley Laboratory, Berkeley, California, 1997), pp. 14 and 26.

3.3.3. Fluorescence microprobe

This is one of the most powerful types of spectromicroscopy, in particular as far as life-science, medical and environmental research is concerned. This technique provides a fast, reliable and reasonably simple way to perform chemical analysis on a microscopic scale. The sophistication level of the analysis can be adjusted to match the specific needs of each application.

The conceptual background is provided by fluorescence spectroscopy (Section 3.2.3). Similar to fluorescence spectroscopy, fluorescence spectromicroscopy can be implemented either in the 'standard' emission-intensity spectroscopic mode or in the excitation-spectroscopy mode (see Figs. 3.40 and 3.41). High lateral resolution is normally achieved by focusing the primary X-ray beam into a small spot.

The applications of fluorescence spectromicroscopy—also widely known as 'fluorescence microprobe analysis'—are widespread and rapidly increasing. Synchrotron sources are quite useful because their intensity increases the signal level and also because they facilitate the task of focusing the primary X-ray beam. The synchrotron fluorescence microprobe has become a standard and powerful analytic tool for many different domains. Figure 3.81 presents two nice examples of this versatile approach.

3.3.4. Photoelectron spectromicroscopy

We have seen (Section 3.2.4) that photoemission-based techniques are excellent probes of the chemical and electronic structure of condensed matter. However, for a long time the lack of lateral resolution constituted a major handicap.

Photoemission without sufficient lateral resolution is of little use in the life sciences and in many domains of chemistry and materials science. Conventional photoemission experiments can only probe areas with a lateral size larger than 0.5–1 mm. This analysis reveals only the average properties of such large areas, and this is unsuitable for studying most life-science specimens.

This problem could not be solved without advanced synchrotron sources. Prior to the high brightness provided by undulators, the signal level of photoemission experiments was not sufficient to achieve good lateral resolution.

The synchrotron facilities commissioned in the late 1980s provided for the first time enough brightness to overcome this obstacle. This triggered a worldwide effort in photoemission spectromicroscopy. As a result, photoemission spectromicroscopy became well established and widely used.

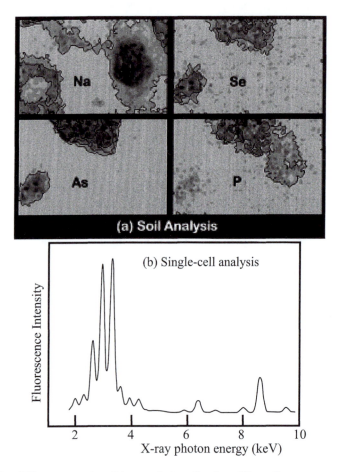

Fig. 3.81 Two different examples of the practical application of X-ray fluorescence analysis on a microscopic scale (fluorescence microprobe): (a) fluorescence intensity maps for different emitted wavelengths show the lateral distribution of different chemical elements in a soil thin section (horizontal size 650 micron); data from: D. G. Strawn, H. E. Doner and S. A. McHugo, *Advanced Light Source Compendium of Users' Abstracts and Technical Reports* 1999 (Lawrence Berkeley Laboratory, Berkeley, California, 1999)). (b) X-ray microfluorescence analysis of single cells: spectrum from a single human ovarian adenocarcinoma cell using a primary focused 14-keV polychromatic X-ray beam [data derived from: S. Bohic, A. Simionovici, A. Snigirev, R. Ortega, G. Deves, D. Hermann and C. G. Schroer, *Appl. Phys. Lett.* **78**, 3544 (2001)].

3.3.4.1. Scanning photoemission spectromicroscopy

Photoemission spectromicroscopy can be implemented with two different approaches: 'scanning spectromicroscopy' and non-scanning or 'electron-imaging spectromicroscopy'—see Fig. 3.82. In the first case, lateral resolution is achieved as for all other scanning approaches by focusing the X-ray beam into a small sample area. We can thus obtain different types of photoemission spectra (EDCs and others) for that specific area.

The position of the focused beam on the sample surface can be changed: normally, this is done by moving the sample with respect to the beam. Local photoemission spectra are thus taken at different sites. Furthermore, by x–y scanning the beam position we can obtain two-dimensional images reproducing the lateral variations of the photoemission intensity over the sample surface.

The detected photoemission intensity can be limited (Fig. 3.83) to photoelectrons of fixed energy corresponding to the excitation of a given core level. The photoelectron intensity 'image' becomes in this way a 'chemical image', showing the local concentration of the chemical element corresponding to that core level.

The photoelectron energy selection can be performed with high resolution. We can thus distinguish not only core-level signals from different elements, but also from the same element in different chemical situations—see again Fig. 3.83.

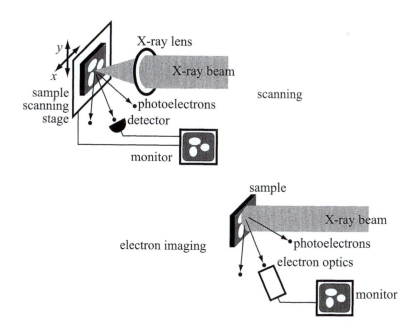

Fig. 3.82 The scanning and non-scanning (electron imaging) modes of photoemission spectromicroscopy.

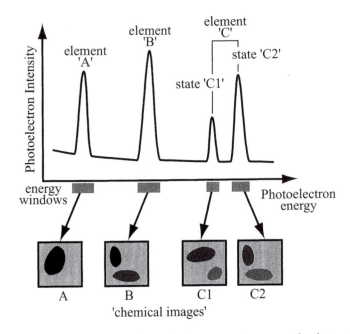

Fig. 3.83 Chemical analysis by scanning photoemission spectromicroscopy: the electron detector can be tuned to different energy windows corresponding to different core levels from different elements (A, B, C). If the energy resolution is sufficient, then the windows can also discriminate between different oxidation states of the same element (C1 and C2). The 'chemical images' reveal the lateral distribution of the corresponding elements or, in the case of element C, of atoms in the two different states C1 and C2.

In this case, the 'chemical images' reveal not only the spatial distribution of the chemical elements, but also that of the oxidation states of each element. This information is of course very valuable for chemical and life-science specimens.

Scanning photoemission spectromicroscopy can be implemented using Schwartzshild objectives, Fresnel zone plates and other focusing devices. Figure 3.84 shows two examples of results. The advantage of the focusing-scanning mode is the possibility of implementing (as long as there is enough signal) on a microscopic scale all the sophisticated photoemission techniques discussed in Section 3.2.4. The main disadvantage is that the time to take an image can be quite long, ranging from seconds to minutes.

3.3.4.2. Electron-imaging photoemission spectromicroscopy

In the electron-imaging mode (see Fig. 3.85), high lateral resolution is achieved by using an electron-optical system to process the emitted photoelectrons. The system is quite similar to an electron microscope with its magnifying electron optics. There is, however, an obvious and important difference: the electrons do not originate from an external electron gun, but are generated by the sample itself via the photoelectric effect.

A photoelectron spectromicroscope based on this approach is called 'PEEM' (PhotoElectron Emission Microscope). In most cases, the operation of a PEEM is based on the partial-yield mode of photoemission spectroscopy (Section 3.2.5)—i.e. the secondary-electron intensity is measured vs the photon energy yielding the surface absorption coefficient. The corresponding spectroscopic features (e.g. absorption edges) reveal the presence of specific chemical elements as well as their chemical status.

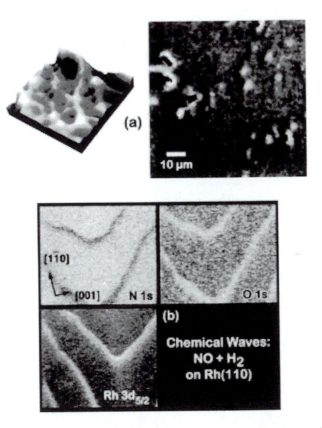

Fig. 3.84 Two examples of results produced by scanning photoemission spectromicroscopy. (a) Images produced by the MAXIMUM spectromicroscope of a neuron network, including on the left a general picture (i.e. an image obtained without energy filtering, shown here after three-dimensional reconstruction), and on the right a 'chemical image' revealing the distribution of the element potassium [data from: G. De Stasio, P. Perfetti, W. Ng, A. K. Ray–Chaudhuri, S. H. Liang, S. Singh, R. K. Cole, Z. Y. Guo, J. Wallace, C. Capasso, F. Cerrina, D. Mercanti, M. T. Ciotti, F. Gozzo and G. Margaritondo, *Phys. Rev.* **E 48**, 1478 (1993)]. (b) Images produced with a zone-plate X-ray lens, showing the lateral distribution of the elements oxygen, nitrogen and rhodium during the phenomenon of 'chemical waves' occurring for the system NO + H_2 on rhodium. Data from: A. Schaak, S. Günther, F. Esch, E. Schütz, M. Hinz, M. Marsi, M. Kiskinova and R. Imbihl, *Phys. Rev. Lett.* **83**, 1882 (1990).

Figure 3.85 shows the general scheme of a PEEM-type system. Instruments of this kind can reach lateral resolution levels of 100–200 Å. They are quite effective for biological and medical studies since they can rapidly deliver large amounts of data. For example, a microchemical analysis can be performed on a large number of cells in a relatively short time.

A PEEM can deliver two types of results: partial-yield intensity images and partial-yield spectra from small sample areas. In the first case, the electron optics projects on the detector plane the lateral distribution of the secondary-photoelectron intensity for a given photon energy. An image of this type carries very valuable chemical information.

In order to understand this point, consider that when the photon energy crosses a core-level X-ray absorption threshold (Section 3.1.1) the partial-yield intensity changes markedly. Suppose that two PEEM images are taken at photon energies immediately below and above a given core-level threshold. Local intensity variations between the two images reveal the presence of the chemical element corresponding to the threshold.

The comparison between the two images produces a 'chemical map' of the surface distribution of the element, carrying information similar to scanning spectromicroscopy images. This approach can be enhanced by performing a pixel-by-pixel digital subtraction of the two images—see Fig. 3.86.

Partial-yield spectra taken with a PEEM on a small sample area reveal the local X-ray absorption coefficient and, through it, the local chemical composition and chemical status. In the most commonly used experimental strategy, the PEEM is first used to take overall images of the sample, and then images and spectra from small areas of specific interest. The most advanced PEEM instruments can also deliver other types of local photoemission spectra—notably EDCs—from small sample areas.

The chemical analysis capabilities of PEEM instruments are frequently exploited for biological and medical research—see the two examples of Fig. 3.87. In addition to the rapid data taking, such instruments offer another advantage with respect to scanning spectromicroscopes: they have limited surface sensitivity.

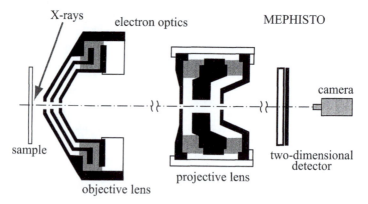

Fig. 3.85 Schematic diagram (not to scale) of the MEPHISTO electron-imaging spectromicroscope, developed by De Stasio *et al.* [see: G. De Stasio, M. Capozi, G. F. Lorusso, P. A. Baudat, T. C. Droubay, P. Perfetti, G. Margaritondo and B. P. Tonner, *Rev. Sci. Instrum.* **69**, 2062 (1998)]. Note the presence of all the elements of an electron-imaging system, as shown in Fig. 3.82.

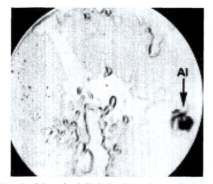

Fig. 3.86 An example of the pixel-by-pixel digital subtraction technique applied to PEEM images. Two images were taken at photon energies immediately above and below the Al 2p X-ray absorption edge. The digital subtraction image shown here emphasizes the aluminum-related signal. In this case, the technique was used to analyze the Al distribution in a rat brain culture specimen after artificial Al contamination—see: G. De Stasio, D. Dunham, B. P. Tonner, D. Mercanti, M. T.Ciotti, A. Angelini, C. Coluzza, P. Perfetti and G. Margaritondo, *Neuroreports* **4**, 1175 (1993).

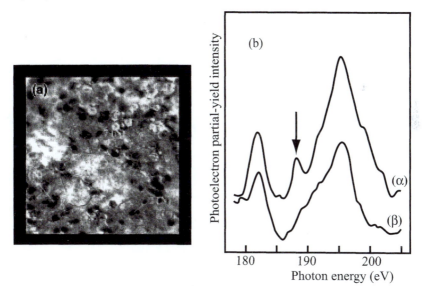

Fig. 3.87 Two examples of biological and medical applications of PEEM spectromicroscopy. (a) A phosphorous distribution map reveals (without staining) the cell nuclei of an ashed human glioblastoma tissue (image size: 100 micron). Data from: B. Gilbert, PhD Thesis, EPFL, Lausanne, 2000. (b) PEEM partial-yield spectra taken in different areas of a human glioblastoma section mounted on a glass slide. The specimen was stained with a boron compound (BSH) to investigate the microchemical aspects of BNCT (boron neutron capture therapy) for brain cancer. Fine spectroscopic analysis reveals a clear BSH-related signal in one area (vertical arrow) but not in the other. Data from: B. Gilbert, L. Perfetti, O. Fauchoux, J. Redondo, P.-A. Baudat, R. Andres, M. Neuman, S. Steen, D. Gabel, D. Mercanti, M. T. Ciotti, P. Perfetti, G. Margaritondo and G. De Stasio, *Phys. Rev.* **E62**, 1110 (2000).

High surface sensitivity is of course desirable when studying surface properties. This is not the case in most biological and medical studies that target instead bulk properties. Excessive surface sensitivity can create problems when scanning spectromicroscopes are used on biological specimens.

This problem is much less severe in the case of PEEM instruments. This point can be understood by considering that the surface sensitivity of photoemission techniques is due to the short photoelectron escape depth—see Fig. 3.55. This figure shows that low-energy secondary electrons have much larger escape depths than primary photoelectrons with larger energies.

Therefore, a partial-yield photoemission spectrum is substantially less surface-sensitive than a conventional EDC spectrum. Similarly, a partial-yield PEEM image or a PEEM spectrum has limited surface sensitivity

Finally, we should note that PEEM instruments can be used to implement on a microscopic scale many different X-ray absorption techniques. These include for example the microscopy version of the EXAFS (extended X-ray absorption fine structure) technique, discussed later.

3.3.5. Microtomography

Many synchrotron-based studies in the life and medical sciences require the analysis of specimens in three dimensions. However, most microscopy techniques project the information from three dimensions to a plane—i.e. to two dimensions. There exist notable exceptions: three-dimensional information is provided by specialized approaches such as 'X-ray tomography' and by its microscopy version, called 'microtomography'.

The working principle of X-ray tomography is similar to the well-known radiology technique 'CAT' (computer-assisted tomography). A CAT scan delivers three-dimensional information about the patient by first taking a set of radiographs with different beam-patient-detector configurations, and then processing the image digital files with a powerful mathematical algorithm.

Similarly, a synchrotron tomography experiment (see Fig. 3.88) is based on the advanced mathematical processing of a sequence of images. Each image reflects the absorption of the X-ray beam by different parts of the specimen, and is recorded with a two-dimensional detector under computer control. Between two images in the sequence, the sample is rotated by a small angle. The procedure is repeated for a large number of images.

As show in Fig. 3.88, a computer-calculated algorithm can then generate from the sequence of raw images different types of processed images. For example, it can yield three-dimensional renditions of the specimen surface.

Furthermore, it can produce sets of 'slice' images of the specimen at different depths and in different directions. In this way, detailed three-dimensional information becomes available in a very useful form.

The high brightness of recent synchrotron sources is necessary to implement this approach on a microscopic scale. The resulting technique of microtomography can produce rather spectacular results—see for example Fig. 3.89 (top). The applications range from biology to materials science and engineering, as shown by Fig. 3.89 (bottom).

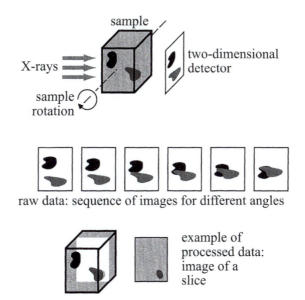

Fig. 3.88 Schematic explanation of the microtomography technique: a sequence of images taken for different (equally spaced) rotation angles of the sample is computer-processed to obtain three-dimensional information, such as images of internal 'slices' of the sample.

3.3.6. X-ray holography

The ultimate X-ray microscope would be an instrument capable of detecting single atoms and molecules and of measuring their positions in three dimensions. Progress has been made in recent years to begin transforming this dream into reality. The approach is based on the extension of holographic techniques from visible light to X-rays.

There are no fundamental conceptual differences between holography in the visible and X-ray holography with atomic resolution—although the practical implementation is quite different. We thus start our discussion with a simple description of visible-light holography.

What is the difference between a normal photographic image and a hologram? In order to clarify this point, note that a normal photographic picture is obtained by recording the intensity of light that has interacted with the imaged object. The most relevant interaction mechanisms are absorption and scattering. Consider the setup of Fig. 3.90 (top): because of absorption, the object is imaged by the photographic plate as a shadow. In Fig. 3.90 (bottom) the situation becomes more interesting: the light is not completely absorbed, and the image reveals different regions inside the object with different levels of absorption.

This second image, therefore, provides information on the internal structure. However, the information is not three-dimensional but projected on a plane and therefore reduced to two dimensions. All normal photographic pictures are in fact two-dimensional images.

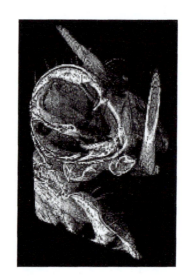

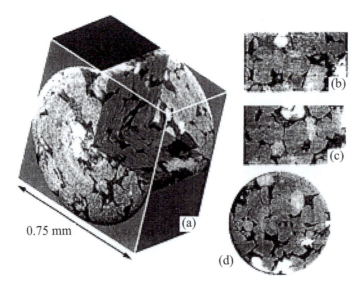

Fig. 3.89 Microtomography results. Top: rendered image from a three-dimensional tomographic reconstruction of a cricket. The image has been digitally cut at a specific plane to show the internal structure of the cricket's head. Data from: Y. Wang, F. De Carlo, D. C. Mancini, I. McNulty, B. Tieman, J. Bresnahan, I. Foster, J. Insley, P. Lane, G. von Laszewski, C. Kesselman, M.-H. Su and M. Thiebaux, *Rev. Sci. Instrum.* **72**, 2062 (2001). Bottom: high-resolution images of a North Sea Brent Sandstone: (a) three-dimensional perspective with cutaway; (b) and (c) slices in two perpendicular directions (x and y); (d) cross section. Data from: M. E. Coles, R. D. Hazlett, E. L. Muegge, K. W. Jones, B. Andrews, B. Dowd, P. Siddons, A. Peskin, P. Spanne and W. E. Soll, Proc. 1996 Society of Petroleum Engineers Annual Technical Conference, Denver, Colorado (SPE 36531); picture courtesy of Keith W. Jones.

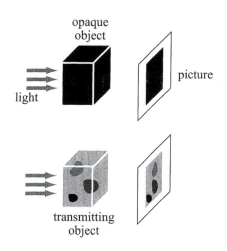

Fig. 3.90 In a normal photographic picture of an opaque (top) or a partially transmitting (bottom) object, the three-dimensional information is reduced to two dimensions.

Holography, on the other hand, provides true three-dimensional information by exploiting the 'phase' of the light waves. The corresponding mechanism of image formation is related to the diffraction phenomena for gratings (Fig. 2.4) and also to phase-contrast imaging (see Inset G). The most important factor is the path difference between different rays of light.

Figure 3.91 illustrates a very simple example of holographic imaging. In Fig. 3.91a, a beam of light illuminates a point-like object (P) that scatters part of it. The light intensity at each point after the object is the combination of non-scattered light waves plus scattered waves. An observer sees the object by detecting the combination of scattered and non-scattered waves and reconstructing the object with his brain.

In Fig. 3.91b, the light after the object is recorded by a coated-glass photographic plate. Assume that the light is monochromatic (i.e. with only one wavelength). The combination of waves at the photographic plate follows the same rules as all other interference phenomena. The combined wave gives maximum intensity if the path difference between the scattered and non-scattered waves is an integer multiple of the wavelength λ. If the path difference is an odd integer multiple of $\lambda/2$, then the combination wave is zero (dark). The result is an imaged interference pattern consisting of concentric circular fringes.

The interference pattern depends of course on the position of the point-like object. Thus, it carries three-dimensional information on the object's position. The problem is to retrieve this three-dimensional information with a simple and reliable procedure

Imagine (Fig. 3.91c) that the photographic plate with diffraction pattern is illuminated with a light beam equivalent to that used to take the image. The darker areas of the pattern absorb the beam more than the lighter areas. As a consequence, the intensity distribution of the wave after passing through the pattern is equivalent to that of the wave of Fig. 3.91a—the observer's brain 'sees' once again the point-like object.

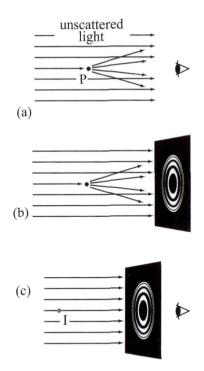

Fig. 3.91 Schematic explanation of a holographic image produced (in the so-called 'Gabor geometry') by a point-like object P. (a) A wave is partially scattered by P, and the superposition of the scattered and non-scattered wave enables an observer to see the object. In (b), the wave after the object is recorded with a photographic plate. The image on the photographic plate is an interference pattern produced by the superposition of scattered and unscattered waves. When the photographic plate with the pattern is illuminated (c) with a wave equivalent to the original one, an observer on the other side detects a wavefront equivalent to that of Fig. 3.91a, and therefore 'sees' an image I of the object.

This intriguing phenomenon can be analyzed based on the theory of Fresnel zone plates (Section 3.3.1.). Consider Fig. 3.92a: the point-like object produces a diffraction pattern that is equivalent to the zone-plate geometry (Fig. 3.70 and eqns 3.6 and 3.7).

What happens when this pattern is illuminated? First, it acts as a Fresnel zone plate concentrating intensity at its focal point. There is, however, another important effect.

As shown in Fig. 3.92b, the path difference condition for constructive wave combinations is also valid for divergent rays. Such divergent rays propagate as if they originated from point (I), which is symmetric with respect to the focal point. An observer detecting the divergent rays thus 'sees' the point-like object—or, more precisely, its virtual image. This is the holographic image of the point-like object.

As shown in Fig. 3.93, the holographic image of a complex object can be analyzed as a combination of point objects that produce zone-plate-like interference patterns. When illuminated, the overall pattern gives the illusion of seeing the combination of point objects—i.e. the complex object in three dimensions.

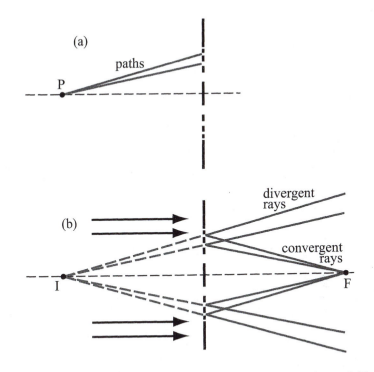

Fig. 3.92 (a) Analysis of the path difference conditions for constructive interference in Fig. 3.91b. Note that the scheme is equivalent (and symmetric) to that of Fig. 3.70 for a zone plate. Therefore, the interference pattern has the same shape as a zone plate. When illuminated (b), the zone-plate-like pattern produces convergent rays that concentrate the intensity on the focal point F. However, the constructive-interference condition for adjacent paths is also valid for the divergent rays. When detected by an observer, such divergent rays produce the image I—which is the holographic image of the point-like object..

This simple discussion explains holographic images taken in the geometry of Figs. 3.91 and 3.92, which is called 'Gabor holography'. Other approaches are used besides the Gabor geometry—see Fig. 3.94. Most important for atomic-scale holographic imaging with X-rays is the 'Fourier holography' of Fig. 3.94b. In this case, the point-object is illuminated by a spherical wave rather than by a plane wave. The source is point-like and located at a distance L from the object.

This brings us to the question of spatial resolution: What are the conditions necessary to 'see' individual atoms in an X-ray hologram? The first condition concerns the wavelength: when λ is too large, one cannot image small objects. Atomic-scale holography thus requires short-wavelength X-rays.

The second issue is the lateral resolution of the detector. Suppose that the object is a molecule with two atoms and that a hologram is taken in the Gabor geometry. From the previous discussion we understand that each atom behaves as a point object producing a zone-plate-like diffraction pattern as in Fig. 3.93.

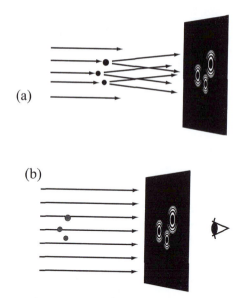

Fig. 3.93 (a) The holographic imaging of a complex object (here, formed by three point objects) can be considered as the combination of holographic patterns produced by individual point objects (see Figs. 3.91 and 3.92). (b) When illuminated, the overall pattern enables the observer to 'see' the image of each point object and therefore the overall complex object.

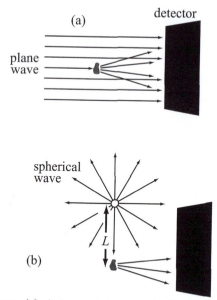

Fig. 3.94 Two geometries used for holographic imaging: (a) the 'Gabor geometry' of Figs. 3.91 and 3.92, based on illumination by a plane wave; (b) the 'Fourier geometry', based on illumination by spherical waves emitted by a point-like source at a distance L from the object. The Fourier geometry is the most important one for atomic-scale X-ray holography.

The diffraction patterns of the two atoms are displaced with respect to each other, and the displacement is determined by the distance δ between the two atoms. Thus, to 'see' the two separate atoms in the hologram the detector must resolve diffraction fringes separated by a distance δ. Since δ is of the order of a few ångströms, this is not technically feasible. Therefore, the Gabor geometry is not suitable for atomic-scale holography.

The situation is different but not much better for Fourier holography (Fig. 3.94b). Theory shows that in this case the approximate diffraction pattern of a point-like object is a series of parallel fringes at a distance $\lambda d/L$ from each other, where d is the source–detector distance and L is the source-object distance. Thus, two point-like atoms separated by a distance δ produce (Fig. 3.95) two overlapping fringe series with periods $\lambda d/L$ and $\lambda d/(L+\delta)$. In order to resolve them, the detector must have lateral resolution better than the difference between the two periods, $\lambda d/L - \lambda d/(L + \delta) = \lambda d\delta/[L(L + \delta)]$, which is approximately equal to $(\lambda d/L^2)\delta$ (as long as δ is much smaller than L).

Thus, the required detector resolution for Fourier holography is no longer δ as for the Gabor geometry, but δ multiplied by the factor $\lambda d/L^2$. By reducing the source–detector distance L and therefore increasing the factor $\lambda d/L^2$, we could in principle compensate for the limited detector resolution and achieve atomic-scale holography.

Unfortunately, this is not true in practice. The Fourier holograms are 'blurred' because the light source is not a point but a finite area. This makes it impossible to obtain atomic-resolution holograms with conventional X-ray sources.

The problem is solved by a clever variation of the Fourier geometry called 'inside source holography'—see Fig. 3.96. A primary X-ray beam is used to stimulate fluorescence from atoms in the object. The fluorescent atoms act as extremely small point-like sources of monochromatic radiation. The imaged objects are the atoms surrounding the emitting ones, and the corresponding source–object distance L is extremely small.

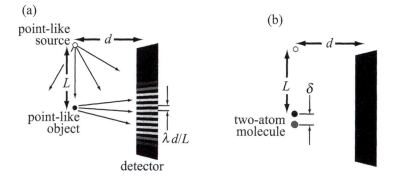

Fig. 3.95 (a) In the Fourier geometry, a point-like object produces a line pattern whose period equals $\lambda d/L$. (b) Two atoms at a distance δ from each other behave like a pair of point objects, producing two overlapping line patterns, whose periods differ by $\approx (\lambda d/L^2)\delta$.

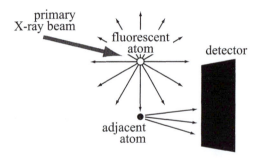

Fig. 3.96 The 'inside source holography': a primary X-ray beam stimulates fluorescence from one atom (the source). The emitted X-rays are partially scattered by a neighboring atom (the object). In reality the approach is used for many similar atoms, e.g. in a crystal.

In order to obtain sufficient signal, this approach must be implemented with a large number of source atoms in equivalent positions with respect to the surrounding object atoms. This is automatically true for atoms in a crystal. However, the technique can also be applied to non-crystalline systems and to 'quasi-crystals' [S. Marchesini, F. Schmithüsen, M. Tegze, G. Faigel, Y. Calvayrac, M. Belakhovsky, J. Chevrier and A. S. Simionovici, *Phys. Rev. Lett.* **85**, 4723 (2000)]. Figure 3.97 shows a rather spectacular example of atomic-resolution holography using the inside source approach. The reconstruction of the hologram was not achieved by direct illumination but by numerically simulating the illumination with a computer.

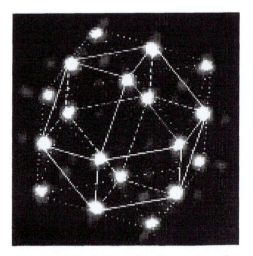

Fig. 3.97 One of the first results of atomic-resolution X-ray holography: reconstructed image of cobalt atoms in a CoO crystal. From: M. Tegze, G. Faigel, S. Marchesini, M. Belakhovsky and A. I. Chumakov, *Phys. Rev. Lett.* **82**, 4847 (1999), copyright 1999 by the American Physical Society. For images of light atoms see also: M. Tegze, G. Faigel *et al.*, *Nature*, **407**, 38 (2000).

Finally, we should note that other approaches related to inside source holography can produce atomic-scale holograms with X-rays. For example, individual atoms are used as detectors in the technique called 'inside detector holography'.

3.4. Extended X-ray absorption fine structure (EXAFS)

Atomic-scale holography could, in principle, solve one of the major problems in chemistry, materials science and the life sciences: identifying the local atomic geometry around a given atomic species. Holography, however, is still far from systematically providing a practical solution.

The EXAFS technique provides instead a simple and reliable way to measure local interatomic distances with good accuracy. Not surprisingly, it has gradually become one of the most widely used synchrotron techniques.

3.4.1. Background: the EXAFS mechanism

The acronym EXAFS refers, as we have seen in Section 3.2.1, to the fine spectral structure found in the absorption coefficient at wavelengths 0.1–0.2 Å beyond each core-level absorption threshold. This structure occurs at smaller wavelengths (larger photon energies) than the NEXAFS—see Fig. 3.30 and the example of Fig. 3.98.

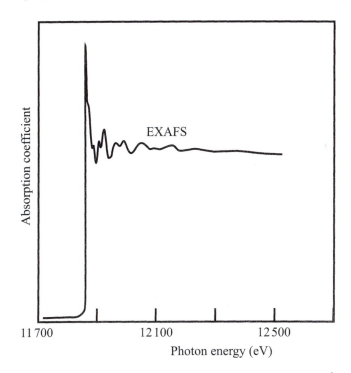

Fig. 3.98 A nice example of long-range EXAFS: the K-edge of implanted As$^+$ in silicon (the structure close to the edge is the NEXAFS). Data derived from: J. L. Allain, J. R. Regnard, A. Bourret, A. Parisini, A. Armigliato, G. Tourison and S. Pizzini, *Phys. Rev.* **B 46**, 9434 (1992).

The mechanism causing the EXAFS is quite simple and explains why EXAFS can be used to measure interatomic distances. We saw (Section 3.1.1, Fig. 3.15) that a core-level absorption threshold corresponds to the annihilation of an X-ray photon whose energy is ceded to a core-level electron. The electron is excited above the photoionization threshold. As shown in Fig. 3.99a, for an isolated atom this implies that the excited electron can leave the atom.

The same conclusion is true for an atom in a molecule or in a solid, but the mechanism is slightly more complicated. The excited electron does not simply behave like a tiny charged particle. According to quantum physics, it also exhibits wave-like properties. As it leaves its atom, it can be imagined as an outgoing spherical wave—see Fig. 3.99b.

The phenomenon can be roughly modeled by the circular waves that are created by throwing a stone into a lake—see Fig. 3.99c. If there are no nearby obstacles, the spherical wave simply propagates away from its source point.

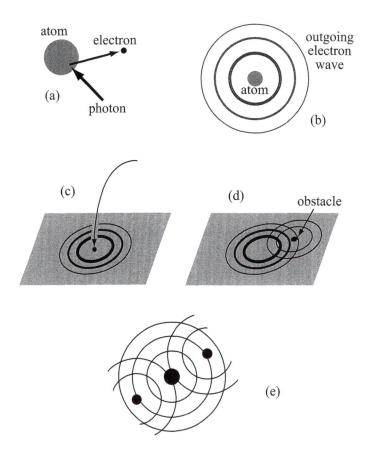

Fig. 3.99 Schematic explanation of the EXAFS mechanism, based on the analogy between electron waves leaving an atom and the waves produced by a stone on the surface of a lake.

Imagine instead (Fig. 3.99d) that a point-like obstacle is encountered by the outgoing wave. The obstacle partly reflects the wave, in particular in the backward direction towards the source point. As a result, the total actual wave is the superposition of the outgoing wave plus the backscattered wave.

Similarly, when an excited electron leaves its atom after absorbing a photon, its spherical quantum wave can be partly backscattered by surrounding atoms. The resulting total wave, as shown in Fig. 3.99e, is the superposition of the outgoing and backscattered waves.

What are the effects of this superposition? The most important one occurs when the two waves cancel each other at the source point: the excitation mechanism becomes difficult, since the electron has no wave-like excited state to reach after absorbing the photon. Even if the cancellation is not complete, the absorption of a photon becomes more unlikely and the absorption coefficient exhibits a minimum. This is the cause of the minima in the EXAFS oscillations—see for example Fig. 3.98.

Figure 3.100 analyzes the condition for the destructive superposition (cancellation) of the outgoing and backscattered waves at the source point. The total round-trip path of the backscattered wave is $2d$, where d is the distance between the source atom and the backscattering atom. Destructive superposition occurs if $2d$ equals an odd-integer multiple of $\lambda_e/2$, where λ_e is the wavelength of the 'electron' wave (not to be confused with the wavelength λ of the absorbed X-ray photon):

$$2d = n(\lambda_e/2) \text{, with } n = 1, 3, 5, \dots \ . \tag{3.10}$$

The implications of eqn 3.10 are very important: if we measure the electron wavelength λ_e, then we can derive from it the local interatomic distance d. This is the foundation of the EXAFS technique.

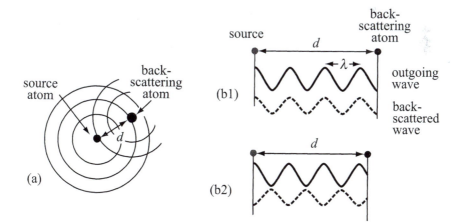

Fig. 3.100 (a) Backscattering by an atom at a distance d from the source atom. (b1) Constructive interference of the outgoing and backscattered waves. (b2) Destructive interference. This simplified picture does not take into account the phase shift (Fig. 3.101).

What is the value of the electron wavelength λ_e? Once again, we must find the answer in quantum physics (see Inset I). The result shows that eqn 3.10 can be written as

$$2d = \frac{nh}{\sqrt{2m(h\nu - E_b)}}, \text{ with } n = 1,3,5,..., \tag{3.11}$$

where E_b is the core-electron binding energy. For each value of the odd-integer number n, eqn 3.11 predicts a minimum in the plot of the absorption coefficient vs the photon energy $h\nu$. The corresponding photon energies are

$$h\nu = \frac{1}{2m}\left(\frac{nh}{2d}\right)^2 + E_b \tag{3.12}$$

(again with $n = 1, 3, 5,...$), and justify the positions of the minima in Fig. 3.99. Note that such minima are not equally spaced along the photonenergy axis: the distance between adjacent minima increases as $h\nu$ increases because n appears squared in eqn 3.12.

Measuring interatomic distances would appear quite simple: once we have found the $h\nu$-values corresponding to the minima in a curve like that of Fig. 3.99, we could use eqn 3.12 to extract the value of d. The real situation is somewhat more complicated.

First of all, in most systems the source atom is surrounded not by one but by several different atoms that backscatter the outgoing electron wave. Fortunately, only atoms that are very close to the source atom significantly contribute to the EXAFS process that otherwise would be messy and useless.

The reason is that an excited electron cannot travel very far in a solid because it rapidly loses its energy. Consequently, the outgoing spherical wave is quickly attenuated: backscattering and EXAFS are only caused by atoms very close to the source atom.

Even so, the EXAFS is still caused by the superposition of backscattered waves by different atoms, and reflects several interatomic distances. Data analysis must separate these effects and retrieve individual interatomic distances—see the next section.

A second cause of complication is illustrated in Fig. 3.101. When the electron wave leaves the source atom it is also slightly shifted. The same thing happens when the wave is backscattered. The total shift (more precisely called the 'phase shift') must be taken into account in the EXAFS analysis. In essence, the effective round-trip path of the backscattered wave is not exactly $2d$ but $2d + x$, where the parameter x describes the effects of the phase shift. Equation 3.11 thus becomes

$$2d + x = \frac{nh}{\sqrt{2m(h\nu - E_b)}} \tag{3.13}$$

In order to derive the value of d from the EXAFS, we must know the value of x.

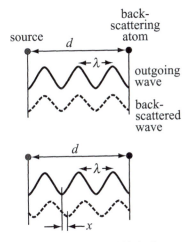

Fig. 3.101 A simple example illustrating the phase shift: in the upper part, the outgoing wave and the backscattered wave combine to produce the maximum oscillation amplitude. In the lower part, the combination is modified because of the phase shift, whose effect is equivalent to changing by x the $2d$ path.

Fortunately, such phase shifts have been measured and calculated. On the other hand, the phase-shift correction is not a trivial task, since x is not a constant but a function of the electron kinetic energy K. Even with these complications (and a few other ones), EXAFS remains a simple and powerful way to measure local interatomic distances.

Note that each EXAFS measurement concerns a specific core-level threshold corresponding to a given atomic species. An EXAFS study, therefore, measures local interatomic distances from a specific type of source atoms, corresponding to a given chemical element. In other words, the EXAFS analysis 'isolates' each chemical element and measures the distances between the corresponding atoms and their neighbors.

This is very important when crucial properties are caused by a minority chemical component with low concentration. Such is the case of many life-science specimens. In the early development of the EXAFS technique its potential importance was in fact stressed by examples like the small difference between hemoglobin and chlorophyll.

Inset I: The wavelength of an electron

Quantum physics shows that the wavelength of an electron in free space corresponds to its momentum mv (where v is the electron speed) according to a law discovered by Louis De Broglie:

$$\lambda_e = \frac{h}{mv},$$
(I1)

where h is of course Planck's constant. In turn, the momentum mv can be derived from the electron kinetic energy, K. Since by definition $K = mv^2/2$, we can write:

$$v = \sqrt{\frac{2K}{m}} \ ,$$

and therefore

$$mv = \sqrt{2mK} \ ,$$

which, inserted in eqn I1, finally gives:

$$\lambda e = \frac{h}{\sqrt{2mK}} \ .$$
(I2)

On the other hand, the energy K of the excited electron is given—see Section 3.2.4, eqn 3.5—by the initial core-level energy (equal to minus the binding energy) plus the photon energy $h\nu$:

$$K = h\nu - E_b$$
(I3)

(this is equivalent to eqn 3.5 except for the absence of the work function). Thus, eqn 3.10 can be written as:

$$2d = \frac{nh}{\sqrt{2m(h\nu - E_b)}} \ , \ \text{with } n = 1,3,5,...,$$

that validates eqn 3.11.

3.4.2. Processing EXAFS data

The EXAFS technique has become quite sophisticated but at the same time it can be implemented almost automatically. The data processing is often based on the so-called 'Fourier analysis'.

This method is also used in other domains of synchrotron research. Therefore, it deserves special attention; in particular, its basic principles discussed here have rather general validity.

The starting point for EXAFS data processing is a raw absorption spectrum plotted as a function of the photon energy $h\nu$—see for example Fig. 3.98. The spectral lineshape after a threshold is generally characterized by a slow decrease. Superimposed onto this decrease we see NEXAFS near the edge and the EXAFS oscillations at larger $h\nu$-values.

The first step in data processing is to extract the EXAFS modulation from the raw data. This is done by first eliminating the $h\nu$ regions below and above the EXAFS. Then, the 'EXAFS modulation' is extracted from the raw spectrum by subtracting a smooth decreasing background.

The result is then divided by the same smooth background to compensate for the progressive decrease in the EXAFS oscillations' intensity. Calling $\alpha(h\nu)$ the raw absorption spectrum and $\alpha_B(h\nu)$ the smooth background, the 'EXAFS modulation function' is thus defined as

$$\chi(h\nu) = \frac{\alpha(h\nu) - \alpha_B(h\nu)}{\alpha_B(h\nu)}. \qquad (3.14)$$

Figures 3.102a and 3.102b show an example of the extraction of the EXAFS modulation function χ from the raw data.

The next step is to convert (using eqns I2 and I3) the variable $h\nu$ into the wavelength λ_e, and then λ_e into a different variable k called the 'wavenumber', defined as

$$k = \frac{2\pi}{\lambda_e}. \qquad (3.15)$$

The reason for this conversion is simple: whereas the minima of the EXAFS modulation plotted vs $h\nu$ are *not* equally spaced, those as a function of k occur, according to eqns 3.10 and 3.15, for

$$k = \frac{n2\pi}{4d} = \frac{n\pi}{2d} \qquad (3.16)$$

(with $n = 1, 3, 5\ldots$), and are equally spaced, with period π/d. The χ-function plotted vs k is thus a regular oscillating function—see Fig. 3.102c and the example of Fig. 3.103.

We can (very roughly) model this oscillating function using a cosine:

$$\chi(k) = A \cos(2kd), \qquad (3.17)$$

where A is a constant. Note that eqn 3.17 gives indeed minima coincident with the k-values of eqn 3.16.

At this point we are ready to apply the Fourier analysis. This consists of calculating the so-called 'Fourier integral' of the function $\chi(k)$, which is defined as

$$F(x) = \int \chi(k) \cos(2kx) dk, \qquad (3.18)$$

with integration extended to all possible values of k. If we use the model χ-function of eqn 3.17, then $F(x)$ is proportional to the integral of the product $\cos(2kd) \cos(2kx)$. Mathematics shows (Inset L) that this integral is zero everywhere except for $x = d$.

Therefore, the Fourier integral $F(x)$ plotted vs x exhibits a sharp peak for $x = d$, as illustrated in Fig. 3.104. The peak position identifies the value of d, the interatomic distance. Thus, the Fourier analysis automatically extracts interatomic distances from the EXAFS modulation.

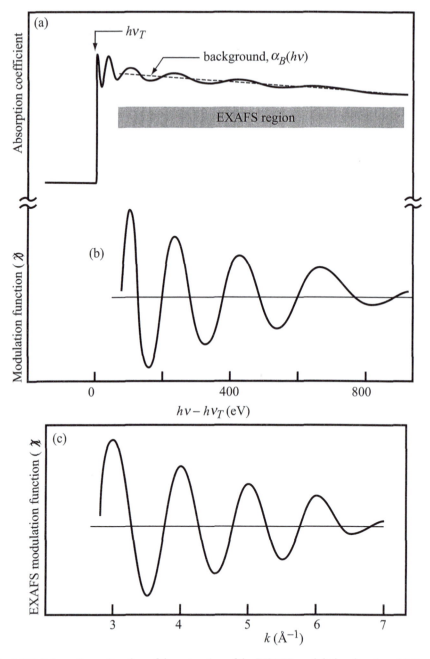

Fig. 3.102 Schematic explanation of the extraction of the EXAFS modulation function. (a) Raw data: a hypothetical absorption spectrum with its EXAFS (the horizontal scale is referred to the absorption threshold). (b) The function $\chi(h\nu)$ (magnified) calculated using eqn 3.14. (c) The function $\chi(k)$ obtained converting $h\nu$ into k—note the regular oscillations with constant period.

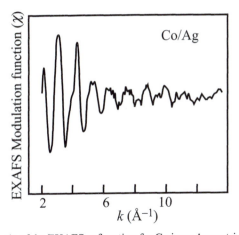

Fig. 3.103 Real example of the EXAFS χ-function for Co in an Ag matrix [from: G. Faraci, A. R. Pennisi, A. Balerna, H. Pattyn, G. E. J. Koops and G. Zhang, *Phys. Rev. Lett.* **86**, 3566 (2001)].

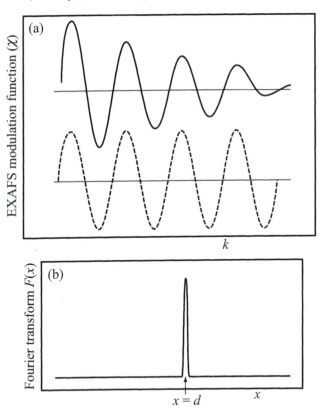

Fig. 3.104 Simplified explanation of Fourier analysis. (a) An EXAFS modulation function χ (top) and its rough model (bottom) based on a cosine function (eqn 3.17). (b) The Fourier integral (eqn 3.18) of χ has a peak that immediately identifies the value d of the interatomic distance.

The advantage of this approach becomes clear if we consider a source atom surrounded not by one but by several different neighboring atoms. The EXAFS modulation is the superposition of the effects of all neighboring atoms. The modulation function $\chi(k)$ thus becomes complicated.

Nevertheless, the Fourier analysis makes its processing very simple. As shown for example in Fig. 3.105, the Fourier integral of the modulation function exhibits several peaks, each identifying the value of one specific local interatomic distance.

Note that the Fourier integral of Fig. 3.105 was calculated for the product $k\chi(k)$ rather than simply for $\chi(k)$. This is a practical way to further compensate the progressive decrease in the EXAFS oscillations. For similar reasons, the Fourier integral can also be calculated for $k^2\chi(k)$ or $k^3\chi(k)$.

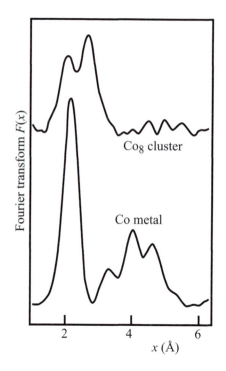

Fig. 3.105 An interesting example of Fourier analysis. The investigated systems were metallic cobalt and small cobalt clusters in a silver matrix. The Fourier integrals of $k\chi(k)$) exhibit different peaks due to different interatomic distances near each cobalt atom. Specifically, for metallic cobalt we see only one strong peak corresponding to the nearest-neighbor Co–Co distance. For the small cobalt clusters in silver, we see instead two strong peaks corresponding to the nearest-neighbor Co–Co and Co–Ag distances. Data derived from: G. Faraci, A. R. Pennisi, A. Balerna, H. Pattyn, G. E. J. Koops and G. Zhang, *Phys. Rev. Lett.* **86**, 3566 (2001); see also, by the same authors, the article in *Physics of Low Dimensional Systems*, J. L. Moran-Lopez Ed. (Kluwer & Plenum, New York, 2001), p. 33.

The identification of the Fourier-integral peaks with interatomic distances is somewhat oversimplified because of several complications. One of them is the phase shift. Another one is the conversion of $h\nu$ into k, requiring an accurate value of the zero point for the electron energy. The zero point, however, is not trivial to evaluate. Furthermore, the calculation of the Fourier integrals (typically performed using imaginary exponentials rather than cosines) is done numerically, and this can cause problems.

In most EXAFS studies, such problems are solved by combining rigorous approaches and empirical solutions. For example, the phase shift can be empirically estimated using a reference compound with the same chemical elements as the investigated sample.

Inset J: Mathematics for the Fourier integrals

The mathematical property underlying the Fourier analysis is that the integral of the product $\cos(a)\cos(b)$ is always zero except for $a = b$. This implies that the integral of $\cos(2kd)\cos(2kx)$ is zero unless $d = x$. Such a property can be easily demonstrated by noting that

$$\cos(a + b) = \cos a \cos b - \sin a \sin b \; ;$$
$$\cos(a - b) = \cos a \cos b + \sin a \sin b \; .$$

Thus, $\cos a \cos b$ can be written as

$$\cos a \cos b = (1/2)[\cos(a + b) + \cos(a - b)] \; . \tag{J1}$$

On the other hand, a cosine is an oscillating function with positive and negative parts, and its integral over the entire range of angles is zero. Thus, both $\cos(a + b)$ and $\cos(a - b)$ give zero integral values, with one exception: if $a = b$, $(a - b) = 0$, and $\cos(a - b) = 1$, which gives a non-zero integral.

3.4.3. Special EXAFS techniques

Many specialized EXAFS techniques have been invented over the past decades. We limit our brief discussion to four special cases: energy-dispersive EXAFS, photon polarization effects, surface EXAFS and micro-EXAFS.

3.4.3.1. Energy-dispersive EXAFS

This technique greatly accelerates the data-taking procedure. The fast data-taking facilitates the real-time EXAFS analysis of systems in rapid evolution.

The data-taking time for standard EXAFS analysis can be quite long, primarily for two reasons. First, the EXAFS is a small modulation superimposed on a strong background and can be detected only if the signal-to-noise ratio is reasonably large. To increase this ratio, we must increase the accumulated signal by lengthening the data-taking time. Second, absorption spectra are normally obtained by scanning $h\nu$. A data point is taken for each $h\nu$-value: this takes time since the number of data points cannot be kept too small without jeopardizing the accuracy in estimating interatomic distances.

The energy-dispersive approach shortens the data-taking time by measuring the absorption at different $h\nu$-values simultaneously rather than in sequence. To see how this is done, we must consider the way X-ray monochromators operate. We saw (Section 2.1.2 and specifically Fig. 2.10) that a crystal monochromator is based on Bragg diffraction. The crystal diffracts in different direction X-rays with different wavelengths and therefore different $h\nu$-values. By selecting with a slit a specific direction, we filter the corresponding $h\nu$-value.

Figure 3.106 shows how Bragg diffraction can be exploited to perform simultaneous measurements for many different $h\nu$-values. The monochromator crystal is bent to focus the X-rays on the sample. Different crystal-to-sample paths correspond to different Bragg diffraction angles and therefore to different $h\nu$-values. On the sample, X-rays with different $h\nu$-values are mixed together. When they reach the detector, they are instead spread apart from each other and therefore detected at different sites.

By using a position-sensitive detector, we can measure the transmitted X-ray intensity simultaneously for all $h\nu$-values. This yields the entire absorption spectrum at once rather than as a sequence of individual steps. Note that the same approach can be used not only for EXAFS but also for all experimental techniques based on X-ray absorption.

Unfortunately, the energy-dispersive approach is not immune to technical problems. For example, the sample and the detector must be highly homogeneous. Such problems notwithstanding, energy-dispersive EXAFS is becoming increasingly popular. Figure 3.107 shows an interesting example of its many applications.

3.4.3.2. Photon polarization effects in the EXAFS analysis
The EXAFS modulation changes with the photon polarization. By studying this effect, we can extract information on the atomic geometry around the source atom. The analysis is quite simple for K-shell absorption edges, which are typically used for EXAFS studies.

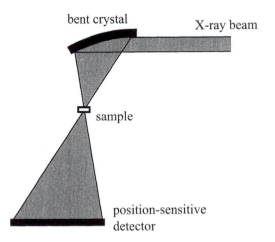

Fig. 3.106 Scheme of an energy-dispersive EXAFS experiment: different photon energies are detected simultaneously, thus reducing the data-taking time and opening the way to real-time EXAFS studies of rapidly evolving systems.

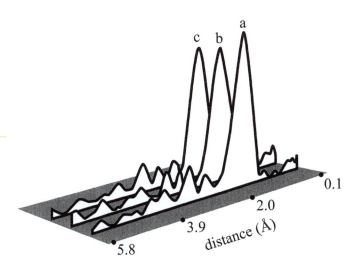

Fig. 3.107 Example of energy-dispersive EXAFS performed as a function of time. The Fourier integrals shown here refer to low loaded $Pt(acac)_2$/H-I SiO_2 catalyst precursors during thermal decomposition and reduction. The curves correspond to different temperatures (a: $T = 30$ C; b: $T = 103$ C; c: $T = 107$ C). Each raw spectrum was taken in 8 s. Data from: S. G. Fiddy, M. A. Newton, A. J. Dent, G. Salvini, J. M. Corker, S. Turin, T. Campbell and J. Evans, *Chem. Commun.* **9**, 851 (1999), reproduced by permission of the Royal Society of Chemistry.

Note that this approach is only feasible when the atoms are in an ordered system like a crystal. The photon polarization in this case is at a well-defined direction with respect to the crystal structure. The EXAFS modulation changes with this direction.

By analyzing the changes, we can identify the orientation of the local chemical bonds around the source-atoms. The same approach can be applied to edges other than the K-edges—but the data analysis becomes much more complicated, and in many cases not practical.

3.4.3.3. Surface EXAFS

In many EXAFS studies the most interesting local structure is that of the surface atoms rather than that of the bulk atoms. This is true, for example, for many catalytic processes and for chemisorption and physisorption phenomena. How can surface interatomic distances be selectively measured with the EXAFS analysis?

This is difficult with standard EXAFS. The same atomic species are typically present both on the surface and in the bulk. The EXAFS contribution from bulk atoms is much stronger than the surface contribution, which is virtually impossible to detect.

Special techniques can be used, however, to filter the surface EXAFS signal thereby eliminating a large fraction of the bulk signal. For example, the absorption coefficient can be measured with the partial-yield technique (Section 3.2.5), which is at least to some extent surface-sensitive.

Alternatively, the surface absorption of X-rays can be measured by detecting the so-called 'Auger electrons' produced by the de-excitation processes after photon absorption. Auger electrons are emitted only from a region very close to the surface and the corresponding EXAFS data are extremely surface-sensitive. Another (somewhat controversial) approach is to detect the intensity of the absorption-generated photoelectron signal.

The surface-sensitive EXAFS produced by these approaches are nicknamed SEXAFS. Figure 3.108 shows an example of a SEXAFS experiment.

3.4.3.4. Micro-EXAFS

Standard EXAFS, like many other synchrotron techniques, does not have high spatial resolution. The analysis is performed on a large sample area and the results correspond to the average properties of this area. This is a severe handicap when interesting properties occur on a more microscopic scale.

Some of the microscopic and spectromicroscopy techniques of Section 3.3 can be used to achieve high lateral resolution in EXAFS. For example, the X-ray beam can be focused on a small sample area. The corresponding technique is known as 'micro-EXAFS'.

3.4.4. Comparison: EXAFS vs crystallography

The EXAFS studies do not typically reach the accuracy of crystallography (see next section) in measuring atomic positions. Why then use it instead of crystallography?

The answer can be found in the differences between the information provided by the two techniques. Crystallography yields global structural information on the system under investigation by using as a probe a large-scale X-ray wave. The EXAFS analysis provides instead local structural information on the immediate neighborhood of the atoms of a specific element by using localized excited-electron waves.

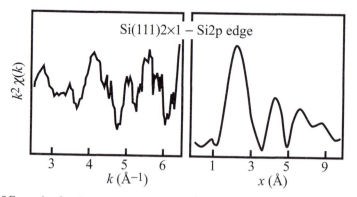

Fig. 3.108 Example of surface-sensitive EXAFS (SEXAFS): analysis of the interatomic distances for atoms forming the surface of a silicon single crystal. The SEXAFS (left-hand) curve was obtained in this case by detecting the photoemitted electron intensity. The right-hand curve shows the Fourier integral with peaks corresponding to surface interatomic distances. Data from: P. S. Mangat, K. M. Choudhari, D. Kilday and G. Margaritondo, *Phys. Rev.* **B44**, 6284 (1991).

Both types of information are very important. They are also largely complementary to each other. The EXAFS analysis becomes very valuable—and often irreplaceable—when it targets system properties caused by a specific chemical component.

3.5. X-ray scattering and crystallography

Identifying the positions of individual atoms in molecules and solids is a key problem in materials science, chemistry and in the life sciences. In fact, the microscopic structure determines many important properties of condensed-matter systems, including a large portion of the functional mechanisms of bio-molecules.

We saw that EXAFS—and potentially holography—are valid tools for atomic-scale structural analysis. However, the (by far) leading microstructural probe is X-ray diffraction and scattering. Some of the related techniques, e.g. protein crystallography, play a key role in the rapid and sustained expansion of synchrotron-based research.

3.5.1. Conceptual background: Fourier transforms

We begin our discussion of X-ray scattering and diffraction with a simple introduction to the basic mechanisms. Their underlying principles can be understood by analyzing once again how we 'see' things using visible light—see Section 3.3.6. Consider Fig. 3.109: an illuminated object scatters light that is then detected by the eye. The lens and the detector system of the eye (Fig. 3.109a) enable the brain to construct the images, i.e. to 'see'.

If the eye is replaced by a photographic camera (Fig. 3.109b), the basic principles are similar. A real image of the object is formed on the photographic plate (the detector) by a suitable lens. Note that the lens processes the information but does not create it: the full information is already contained in the scattered light.

Suppose then eliminating the lens so that the scattered light is directly recorded by the detector (Fig. 3.109c). The detector no longer produces a picture corresponding to the image of the object. Can we nevertheless use the recorded scattered light to reconstruct the object's structure? Our previous discussion of holography suggests that the answer is positive.

3.5.1.1. The concept of the Fourier transform

The link between the scattered light and the structure of the scattering object is provided by a powerful mathematical tool called the 'Fourier transform'—see Fig. 3.110. The Fourier integral used in Section 3.4 for EXAFS is an example of the Fourier transform. In that case, the Fourier transform linked the EXAFS oscillations and the atomic positions around the source atom.

A similar concept applies to scattered light (or, more precisely, to scattered X-rays) and atomic positions in the scattering object. Imagine for example (Fig. 3.111a) that the object is a very small sphere that can be virtually considered as a point. The scattered light is nearly homogeneously distributed over the 'forward' directions: the recorded intensity on the photographic plate is a slowly decreasing pattern with circular symmetry.

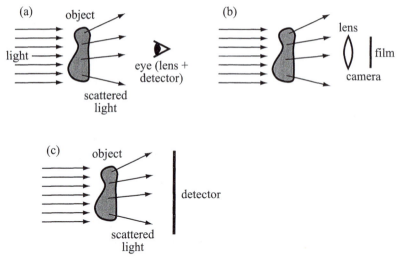

Fig. 3.109 (a) and (b) The mechanisms for seeing an object with the eye or for taking its picture with a camera are similar. In both cases the scattered light is processed by a lens and then detected. (c) On the other hand, the scattered light can also be directly detected without passing through a lens system: from the intensity pattern we can still derive information on the object's structure.

Similar to the case of EXAFS, by applying an appropriate Fourier transform to the scattered wave we can retrieve the structure and position of the point-like object. We can then say that, roughly speaking, *the object is the Fourier transform of the wave*.

We must now slightly complicate the situation and consider a scattering object formed by two point-like spheres at a distance δ from each other—see Fig. 3.111b. The waves scattered by the two spheres combine according to the usual path-difference rules. Specifically, a path difference equal to an integer multiple of the wavelength gives maximum intensity. The corresponding intensity pattern is also shown in Fig. 3.111b.

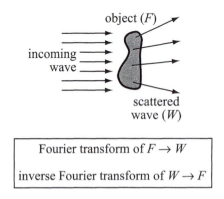

Fourier transform of $F \to W$

inverse Fourier transform of $W \to F$

Fig. 3.110 The scattered wave W and the structure F of the scattering object are mathematically linked by an operation called the 'Fourier transform' that operates on F to give W. The inverse link between F and W is also a type of Fourier transform.

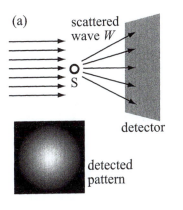

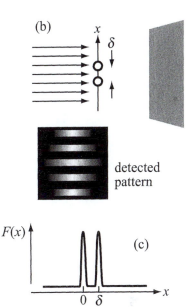

Fig. 3.111 Light scattering by simple objects: (a) scattering by a point-like sphere S; the corresponding detected intensity pattern is circularly symmetric; (b) scattering by two point-like spheres at a distance δ; (c) the two-peaked position function $F(x)$ describing the structure of the two-point scattering object.

This intensity pattern once again reflects the overall scattered wave. By applying a proper Fourier transform, we can retrieve from this wave the structure of the two-sphere object. Such a two-point structure is described by the two-peaked object position function $F(x)$ of Fig. 3.111c. Again calling W the scattered wave, we can say that

$$F(x) = \text{the Fourier transform of } W. \tag{3.19}$$

Inset K discusses the specific nature of the Fourier transform linking W and F.

The link expressed by eqn 3.19 can be inverted: *the inverse of a Fourier transform is also a type of Fourier transform.* We can then say that

$$W = \text{the inverse Fourier transform of } F(x). \tag{3.20}$$

Equations 3.19 and 3.20 constitute the backbone of X-ray structural analysis. They confirm that the scattered wave contains information about the object structure and show that data processing based on Fourier transforms can retrieve this information.

Figure 3.112 schematically shows two examples of Fourier transforms linking different scattering objects and the corresponding waves, revealed by their intensity patterns. Note that the link established by Fourier transforms is valid not only for one-dimensional objects like the two-point system discussed above but also for two-dimensional (Fig. 3.112) and three-dimensional objects.

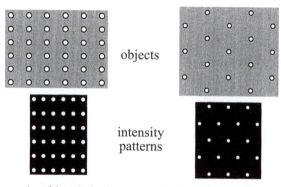

objects

intensity
patterns

Fig. 3.112 Two examples of the relation between scattering objects and scattered waves described by Fourier transforms. Here, the waves scattered by two different two-dimensional crystals (top) are revealed by detecting the corresponding intensity patterns.

Inset K: Fourier transforms for X-ray scattering and crystallography

The use of Fourier transforms for X-ray scattering and diffraction is more effectively treated by using imaginary exponentials instead of the cosine (or sine) functions utilized in Inset J to discuss EXAFS. Note that sines, cosines and imaginary exponentials are linked by Euler's equation:

$$\exp(i\alpha) = \cos\alpha + i\sin\beta. \tag{K1}$$

In order to introduce the basic notions about Fourier transforms, we can use a simple case like Fig. 3.111b. The scattered wave geometry is illustrated by Fig. K-1. The angle θ that defines the direction of the two rays is $\approx y/L$ (this is an approximation, valid as long as the angle is small; we also assume that δ is so small that the angle θ is nearly the same for both rays). The path difference between the two rays is then $\approx \delta\theta \approx \delta(y/L)$.

Each wave can be expressed as an oscillating function of the variable time, t. Assume then that wave A at the detector point y is proportional to the oscillating function $\exp(i\omega t)$. The oscillation of wave B at the same detector point is advanced in time because of the shorter path with respect to wave A. We can thus derive wave B by using the following argument. First, when the path difference is $\lambda/2$, we must have destructive interference. This requires the arguments of the exponential functions for the two waves to be different by $i\pi$. Therefore, a generic path difference $\delta(y/L)$ must produce a difference between the two arguments equal to

$$i\pi\frac{\delta(y/L)}{y/L} = \frac{2i\pi y\delta}{L\lambda}.$$

Wave B must then be proportional to the exponential function:

$$\exp\left[i\left(\omega t - \frac{2\pi y\delta}{L\lambda}\right)\right] = \exp(i\omega)\exp\left(-i\frac{2\pi y\delta}{L\lambda}\right).$$

The total scattered wave W is the sum of waves A and B, therefore

$$W(y) \text{ proportional to } \left[\exp(i\omega t) + \exp(i\omega t)\exp\left(-i\frac{2\pi y\delta}{L\lambda}\right)\right]. \tag{K2}$$

We can simplify this equation by defining a new variable k as:

$$k = 2\pi y/(L\lambda), \tag{K3}$$

so that eqn K2 becomes

$$W(k) \text{ proportional to } [\exp(i\omega t) + \exp(i\omega t)\exp(-ik\delta)]$$
$$= \exp(i\omega t)\,[1 + \exp(-ik\delta)],$$

and therefore

$$W(k) \text{ proportional to } [1 + \exp(-ik\delta)]. \tag{K4}$$

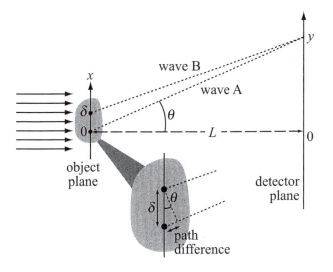

Fig. K-1

Assume that during an X-ray scattering experiment we measure the scattered wave W described by eqn K4. We can retrieve the object's position function $F(x)$ by calculating the Fourier transform of $W(k)$. The definition of the Fourier transform valid in this case is reminiscent of that of eqn 3.18 (for EXAFS), but the cosine function is replaced by an imaginary exponential:

$$F(x) \text{ proportional to } \int W(k)\exp(ikx)dk . \tag{K5}$$

Considering the form of $W(k)$ (eqn K4) and Euler's equation (eqn K1), we can write eqn K5 as

$$
\begin{aligned}
F(x) \text{ proportional to } & \int [1 + \exp(-ik\delta)]\exp(ikx)dk \\
= & \int [\exp(ikx) + \exp(-ik\delta)\exp(ikx)]dk \\
= & \int [\cos(kx) + i\sin(kx) + \cos(k\delta)\cos(kx) - i\sin(k\delta)\cos(kx) \\
& + i\cos(k\delta)\sin(kx) + \sin(k\delta)\sin(kx)]dk .
\end{aligned}
$$

The integral of the term $\cos(kx)$ is zero except for $x = 0$. The terms $\cos(k\delta)\cos(kx)$ and $\sin(k\delta)\cos(kx)$ give zero integrals except for $x = \delta$. The other terms always give zero integrals. Thus, $F(x)$ is zero except for $x = 0$ and $x = \delta$—and the Fourier transform of the scattered wave $W(k)$ does indeed reproduce the two-peaked object structure $F(x)$.

As to the inverse link between $F(x)$ and $W(k)$, we must use the inverse Fourier transform defined as

$$W(k) \text{ proportional to } \int F(x)\exp(-ikx)dx . \tag{K6}$$

Assume for simplicity that the two-peaked function $F(x)$ is zero everywhere except for the two peaks: $x = 0$ and $x = \delta$. The only contributions to the integral of eqn K6 are $\exp(0) = 1$ and $\exp(-ik\delta)$, so that:

$$W(k) \text{ proportional to } 1 + \exp(-ik\delta) .$$

This is equivalent to eqn K4. Thus, the inverse Fourier transform does indeed convert $F(x)$ into $W(k)$.

3.5.1.2. Why X-rays?

Our discussion of scattering has so far been generically based on 'light'. Atomic-scale structural analysis, however, specifically requires short-wavelength waves, i.e. X-rays. The reason can be easily understood. Suppose that the object in Fig. 3.111b is a biatomic molecule and that the two small spheres are its two atoms. The distance δ is then of the order of a few ångström.

According to the rules for destructive interference, the first intensity minimum occurs (see again Inset K) when the path difference $\delta(y/L)$ equals $\lambda/2$, which gives

$y = (L/2)(\lambda/\delta)$.

Suppose that the scattered wave is visible light rather than X-rays. Typical values $\lambda = 5000$ ångström for visible light and $L = 1$ m would give a distance y of several hundred meters to more than one kilometer. Thus, to catch the intensity minimum we would need a detector size of at least several hundred meters! This is entirely unrealistic: actual detectors are much smaller. However, without detecting at least the first intensity minimum we would miss the structural information carried by the wave.

To avoid this problem, the ratio λ/δ cannot be too much larger than unity. Thus, the wavelength must be in the range 0.1–10 Å, typical of X-rays. This illustrates why structural investigations require X-rays.

3.5.1.3. Small angles vs large angles

A realistic experiment detects scattered X-ray waves not in all directions but only over a limited angular range, as shown in Fig. 3.113. What is the effect of limiting the detected angles? We can find the answer by once again using the properties of the Fourier transforms.

Consider for example the small molecule of Fig. 3.114, composed of three atoms on a line. The figure also shows the position function $F(x)$ and the real part of the scattered wave $W(k)$, corresponding to the Fourier transform of $F(x)$.

Suppose now that the experiment detects only waves scattered at small angles, as shown in Fig. 3.114. When we retrieve (by Fourier transform) the object structure $F(x)$ from the measured waves, we obtain the overall shape of the object but not its fine details. If we expand the detection to larger angles, we reveal finer structural details.

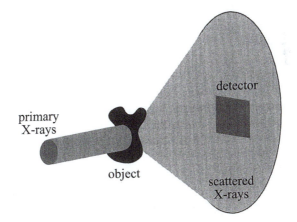

Fig. 3.113 In an actual experiment, X-rays are scattered in a very broad angular range (in fact, in all directions). However, the detector collects X-rays only in a limited portion of this range.

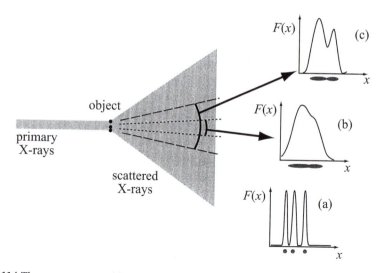

Fig. 3.114 The consequence of limited angular collection is loss of information—as illustrated by this simple example of X-ray scattering by an object formed by three point-like components. The position function $F(x)$ along the x-axis is shown in (a), and consists of three sharp peaks. The objective of the experiment is to retrieve $F(x)$ from scattering data. However, if the detection is limited to small angles (b) the detailed structure of $F(x)$ cannot be retrieved and only the general features of $F(x)$ are obtained. As the angular collection range becomes larger (c), finer details of $F(x)$ are revealed.

This simple example leads us to an important discovery. The general information about the object's shape is contained in the waves scattered at small angles, whereas to reveal the fine microscopic structure we must detect scattered waves at larger angles. This illustrates the difference between the technique of 'Small-angle X-ray Scattering (SAXS)' (scattering angle < 1–2 degrees) that investigates structural information on the scale of 10 Å or more, and large-angle X-ray scattering.

Why is the fine structure related to large-angle scattering? Consider for example Bragg's law, eqn 2.3 [$2d \sin \theta_d = \lambda$]: $\sin \theta_d$ is proportional to $1/d$ and thus it increases as d decreases. To reveal a small crystal periodicity, we must look at large scattering angles. A similar analysis applies to non-crystalline systems.

3.5.1.4. Role of synchrotron light

Synchrotron sources are very important in X-ray scattering and diffraction for several different reasons. Some specific structural techniques strictly require the performances of synchrotron sources—most notably the Multiwavelength Anomalous Diffraction (MAD) method to be discussed in Section 3.5.6. In general, a synchrotron source enhances the effectiveness and accuracy of virtually all the structural analysis techniques.

Consider, for example, the fact that our previous discussion assumed X-rays of only one wavelength and reaching the specimen along a well-defined direction. Neither the first nor the second hypotheses is automatically verified.

A conventional X-ray source emits instead a wide band of wavelengths over a broad range of directions. On the other hand, a 'blurred' wavelength and/or a not-so-well-defined direction degrade the accuracy of the structural analysis.

To avoid this problem, we must use monochromatic (narrow wavelength band) and collimated X-rays. Monochromaticity requires filtering wide-band radiation with a monochromatizing device. This decreases the flux thereby creating serious problems, which are alleviated by the high initial brightness of a synchrotron source. As to the direction, synchrotron X-rays are naturally collimated whereas those emitted by conventional sources must be collimated using a pinhole system—with a substantial waste of flux.

3.5.2. Small-angle X-ray scattering (SAXS)

We saw that SAXS experiments neglect the fine structural information carried by X-rays scattered at large angles, and analyze instead general structural features on a length scale of 10 Å or more. This information is extremely valuable for many different systems—such as gels, liquid crystals, biopolymers and muscle tissues. SAXS is very important in the study of amorphous materials in general since, unlike crystallography, it does not require crystals with long-range order. Highly intense synchrotron sources can study the structure of evolving systems by real-time SAXS analysis.

Many SAXS experiments are performed under the 'Guinier approximation'. Such studies deliver for example information on the so-called 'radius of gyration' R_g, a parameter characterizing the overall structure of the object. In order to understand this notion, consider for example a biological molecule (Fig. 3.115). Its internal structure can be very complicated. However, seen from far away the molecule looks to a first approximation like a particle. We can thus model it (Fig. 3.115) as a small sphere whose radius is the 'radius of gyration' R_g of the molecule.

In the Guinier approximation, the scattered X-ray intensity $I(\theta)$ at the deviation angle θ is linked to the R_g-parameter by the equation

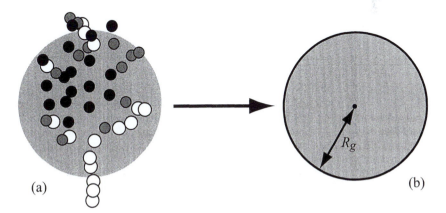

(a) (b)

Fig. 3.115 A biological molecule can be rather complicated (a). However, seen from a large distance (b) it 'looks like' a simple sphere, whose radius is R_g, the radius of gyration.

$$I(\theta) \approx I_0 \left[1 - \frac{1}{3} \left(\frac{R_g 2\pi\theta}{\lambda} \right)^2 \right],$$

(3.21)

where I_0 is a constant. The justification of eqn 3.21 is discussed by Inset L. Its usefulness is clear: measurements of the intensity vs the angle θ directly yield the R_g-value.

The applications of SAXS extend well beyond the measurements of R_g in the Guinier approximation. Figure 3.116 shows two interesting examples of such applications. Note, in particular, the time-resolved dynamic study of the second case.

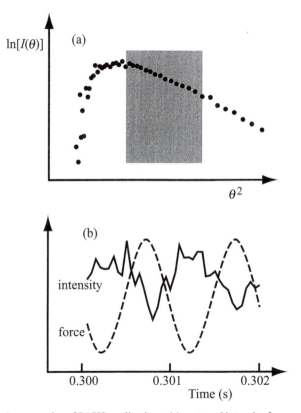

Fig. 3.116 Two nice examples of SAXS applications: (a) scattered intensity for aqueous suspensions of synthetic tyrosine-derived melanin with a copper-to-melanin molar ratio of 0.2. The shaded area shows the region for which the Guinier approximation works reasonably well (data from: R. D. Glickman, J. M. Gallas, S. L. Jacques, B. A. Rockwell D. K. and Sardar, in *Saratov Fall Meeting 2000: Optical Technologies in Biophysics and Medicine II*, V. V. Tuchin, Ed., *Proc. SPIE* **4241**, 112 (2001); (b) time dependence of the M3 (axial cross-bridge periodicity) diffraction spot intensity (solid line) and force (dashed line) for the living skeletal muscle cells of a frog, revealing the structural aspects of the muscle mechanism during sinusoidal length oscillations. Data from: H. Amenitsch, C. C. Ashley, M.A. Bagni, S. Bernstorff, G. Cecchi, B. Colombini and P. J. Griffiths, *ELETTRA News* **26** (1998); also see: M. A. Bagni, G. Cecchi, B. Colombini, P. J. Griffiths, C. C. Ashley, H. Amenitsch and S. Bernstorff, *J. Physiol. London* **533**, 135 (2001).

Inset L: The Guinier approximation

Equation 3.21 is a consequence of the Fourier transform that links the wave W and the object position function F. Consider, for example, the two-point scattering objects of Fig. 3.111b. Equation K-4 shows that the scattered wave $W(k)$ is proportional to $1 + \exp(-ik\delta) = [1 + \cos(k\delta)] - i\sin(k\delta)$.

The theory of electromagnetic waves shows that the intensity $I(\theta)$ is proportional to the square $|W|^2$ of the wave $W(k)$. Since $W(k)$ is a complex quantity with real and imaginary parts, its square must be calculated using the standard rules for complex numbers.

By definition, the square of a complex number equals the square of its real part plus the square of its imaginary part. Therefore, the intensity $I(\theta)$ is proportional to

$$
\begin{aligned}
|W(k)|^2 &= [1 + \cos(k\delta)]^2 + [-\sin(k\delta)]^2 \\
&= 1 + 2\cos(k\delta) + \cos^2(k\delta) + \sin^2(k\delta) \\
&= 1 + 2\cos(k\delta) + 1 = 2[1 + \cos(k\delta)] \ .
\end{aligned}
$$

On the other hand, for small angles corresponding to small values of the parameter $k = 2\pi z/(L\lambda)$, we have $z/L \approx \theta$ and $k \approx 2\pi\theta/\lambda$. We can then write:

$$I(\theta) \text{ proportional to } 1 + \cos(2\pi\theta\delta/\lambda) \ . \tag{L1}$$

Again for small angles, $\cos(2\pi\theta\delta/\lambda) \approx 1 - (1/2)(2\pi\theta\delta/\lambda)^2$, and we obtain:

$$
\begin{aligned}
I(\theta) &\text{ proportional to } 2 - (1/2)(2\pi\theta\delta/\lambda)^2 \\
&\text{ proportional to } 1 - (\delta/2^2)(2\pi\theta/\lambda)^2.
\end{aligned} \tag{L2}
$$

This result corresponds to the Guinier approximation of eqn 3.21 if we assume that $R_g^2/3 = \delta^2/4$ and therefore $R_g = (\sqrt{3}/2)\delta$.

The reader familiar with classical mechanics should recognize this result. In fact, by modeling the two-particle system as a sphere, the moment of inertia of the sphere would correspond to a radius $\approx (\sqrt{2.5}/2)\delta$, quite close to the value $R_g = (\sqrt{3}/2)\delta$.

3.5.3. X-ray diffraction by periodic structures: general features

Many X-ray structural investigations concern crystals or other periodic structures. Such investigations can identify both the overall crystal structure and the positions of the individual atoms within such a structure.

This leads in particular to an effective strategy for analyzing the atomic structure of biological macromolecules (e.g. proteins). The molecules can be artificially arranged in crystals and analyzed with X-ray diffraction. Note that the targeted information in this case is not the overall structure of the artificial crystal but the atomic structure of the individual molecules within it.

Why analyze a crystal rather than one single molecule? The main reason is being able to work simultaneously on many equivalent molecules. Consider radiation damage: the X-rays can modify a molecule generating spurious results. In a crystal with very many molecules, the damage is spread among them and its effects are limited.

3.5.3.1. X-ray diffraction by a crystal: Bragg spots

Bragg diffraction by a crystal was discussed in Sect. 2.1.2 when we treated the working principles of crystal monochromators. The basis is Bragg's law (eqn 2.3 and Fig. 2.10) which predicts X-ray diffraction along specific and well-defined directions. Each one of these directions would correspond to a bright 'Bragg spot' in a luminescent two-dimensional X-ray detector, and is related to a specific family of crystal atomic planes.

Bragg's law [$2d \sin \theta_d = \lambda$] also explains how the Bragg spots can yield information on the overall crystal structure. This law relates in fact the direction of each Bragg-diffracted beam both to the orientation of the corresponding family of crystal planes and to the distance d between adjacent planes.

The treatment of the Bragg spots can be broadened by considering the properties of the Fourier transforms of periodic functions. Once again, the scattered wave W is the Fourier transform of the position function $F(x)$ that describes the positions of individual atoms (and of their electrons) in the crystal. Since the crystal is periodic, the F-function is also periodic—see for example Figs. 3.117a and 3.117b. Its Fourier transform W is again a periodic function, as shown by the examples of Figs. 3.112 and 3.117c. Thus, diffraction by a crystal produces a periodic array of Bragg spots on the detector.

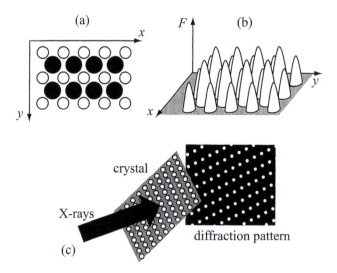

Fig. 3.117 The notion of Fourier transforms applied to a periodic (two-dimensional) crystal. The overall crystal periodicity (a) corresponds to a position function F (b) of similar symmetry. (c) A Fourier transform reflects the symmetry of the transformed function: as a consequence, the diffraction pattern of a periodic crystal is also periodic, although with a different geometry.

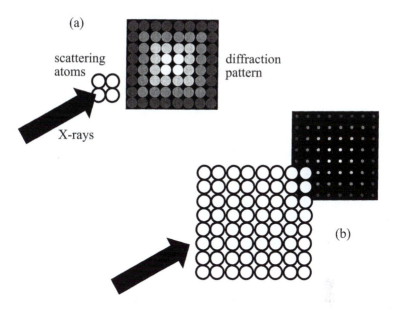

Fig. 3.118 The sharpness of the Bragg spots depends on the number of atomic planes in the scattering crystal: (a) a few atomic planes give broad spots, whereas (b) a large number of atomic planes produces sharp spots.

Note that these arguments explain the periodicity of the Bragg spots but not their sharpness. This is a consequence of the large number of atomic planes present in a crystal—see Inset M. As schematically illustrated in Fig. 3.118, the spot width decreases rapidly as the number of atomic planes increases. A typical crystal contains very many atoms and atomic planes, thus its Bragg spots are very sharp.

The periodic positions of the Bragg spots on the detector provide information on the overall structure of the crystal. But what about the positions of individual atoms? Consider Fig. 3.119: we can ideally 'construct' a crystal by combining individual 'building blocks' formed by atoms (and by their electrons). Can the diffracted X-rays reveal the structure of each building block (or 'unit cell')? This is a key point: diffraction theory shows that *the intensity of individual Bragg spots yields information on the atomic positions within the unit cell*.

3.5.3.2. The crystal structure factor

In order to understand this last statement, we must analyze in detail the scattering mechanism. First of all, the scattering is actually produced not by the atoms but by their electrons. Therefore, X-ray scattering by a crystal reflects the overall electronic charge distribution inside the crystal.

Consider (Fig. 3.119) the electronic charge distribution within each unit cell. In order to 'construct' the electronic charge distribution of the entire crystal we must arrange the unit cells in a periodic fashion according to the crystal structure.

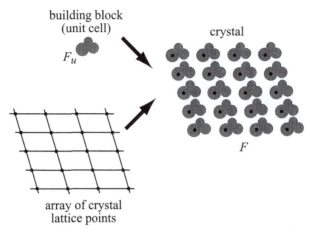

Fig. 3.119 The overall electronic charge distribution F of a crystal is the combination (or, better, the convolution) of the electronic structure F_u of a unit cell with the array (L) of lattice points that characterizes the periodicity of the crystal.

From a formal point of view, this procedure starts with the function F_u (similar to the object position function of Section 3.5.1) that describes the charge distribution within an individual unit cell. Then, one F_u-function is placed at each one of the lattice points of the crystal—see again Fig. 3.119. This procedure gives the overall electronic charge distribution function F.

Thus, the F-function is the combination of the unit-cell function F_u with the array of lattice points. Calling L the function describing the lattice points, F is called the *convolution* of F_u with L.

The scattered X-ray wave W is the (inverse) Fourier transform of the overall charge distribution function F. This is equivalent to the Fourier transform of the convolution of F_u with L.

Mathematics shows that the Fourier transform of the convolution of two functions is the product of the Fourier transforms of the two functions. Thus (see Fig. 3.120 and Inset M), the wave W is the product of the Fourier transforms of L and F_u.

What is the specific role of each of these two Fourier transforms? The L-function, which reflects the crystal symmetry—corresponds to the periodic arrangement of the Bragg spots.

On the other hand, the *intensity* of each Bragg spot is determined by the corresponding wave amplitude. In turn, this is proportional to the Fourier transform of F_u. Thus, the Bragg spot intensity is indeed determined by the function F_u that reflects the individual atomic positions in each unit cell, as stated above.

In crystallography, the Fourier transform of F_u is the so-called 'crystal structure factor', f. A crystallography experiment measures the scattered X-ray intensity thereby obtaining information both on L and on f. Specifically, the overall periodicity of the Bragg spots carries information on L and therefore on the general crystal structure. The intensity of the individual Bragg spots carries instead information on the crystal structure factor f—and therefore on the atomic arrangement of each unit cell.

These arguments clarify in particular how X-ray crystallography identifies the structure of macromolecules—see again Fig. 3.120. First, many macromolecules are artificially arranged in an ordered crystal. Each macromolecule constitutes a single unit cell, and its atomic structure determines the electronic charge distribution (F_u-function) of the unit cell.

X-ray scattering by the macromolecular crystal produces a pattern of Bragg spots. By measuring the intensity of the Bragg spots, we obtain the crystal structure factor f. From f, we can derive the F_u-function which is its Fourier transform. The F_u-function describes the detailed atomic structure of a molecule. In summary, by measuring the intensity of Bragg spots we derive the arrangement of atoms in a molecule.

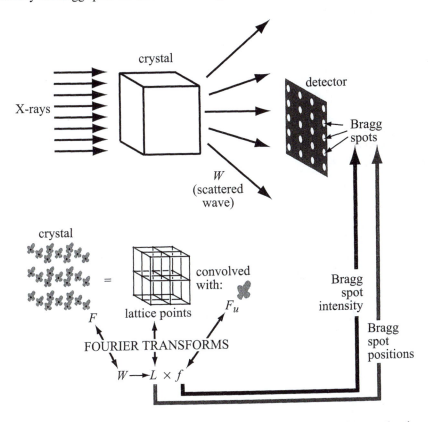

Fig. 3.120 Schematic explanation of macromolecular crystallography. Many molecules are artificially arranged in a crystal. X-ray diffraction by the crystal produces Bragg spots. The electronic charge distribution F in the crystal is the convolution of the lattice points (L) with the charge distribution F_u of a molecule (a unit cell). Owing to the Fourier transform properties, the diffracted wave corresponds to the product of the Fourier transform of the lattice-point function L times the Fourier transform of F_u—called the *crystal structure factor f*. L determines the positions of the Bragg spots whereas f determines their intensity levels. From the Bragg spots we can thus retrieve both L and f, and from f the crystal structure and F_u—corresponding to the atomic structure of a single molecule.

Inset M: Fourier transforms of periodic functions

The electronic charge distribution for a crystal is a periodic function F reflecting the crystal's symmetry. The Fourier transforms of periodic functions have specific properties that influence the scattering of X-rays by crystals. For simplicity, we discuss such properties in one dimension.

Consider the one-dimensional periodic function $L(x)$ of Fig. M-1a, consisting of a series of N sharp peaks at $x = d, 2d, 3d,\ldots$. According to eqn K6, its Fourier transform is defined as

$$W(k) \text{ proportional to } \int F(x)\exp(-ikx)dx ,$$

which in this case becomes

$$W(k) \text{ proportional to } \sum_{n=1}^{N} \exp(-inkδ) .$$

This is a geometric series, whose sum is given by

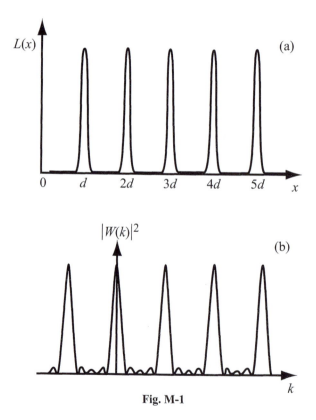

Fig. M-1

$$\sum_{n=1}^{N} \exp(-ink\delta) = \frac{1 - \exp(-iNk\delta)}{1 - \exp(-ik\delta)} \ ;$$

it is easy to verify that the (complex-number) square of this quantity (by definition, the square of its real part plus the square of its imaginary part) is proportional to

$$\left[\frac{\sin\left(-\dfrac{Nk\delta}{2}\right)}{\sin\left(-\dfrac{k\delta}{2}\right)} \right]^2 ,$$

so that the scattered wave $W(k)$ produces an intensity pattern:

$$|W(k)|^2 \text{ proportional to } \left[\frac{\sin\left(-\dfrac{Nk\delta}{2}\right)}{\sin\left(-\dfrac{k\delta}{2}\right)} \right]^2 . \tag{M1}$$

This function is plotted in Fig. M-1b for $N = 5$. The peaks of Fig. M-1a could represent the lattice points of a one-dimensional crystal. In that case, the peaks in the intensity pattern function $|W(k)|^2$ of Fig. M-1b would be the 'Bragg spots'.

It can be shown that eqn M1 also implies that the sharpness of the peaks and therefore the narrowness of the Bragg spots increases with N, the number of periods. This is the property illustrated by Fig. 3.118.

The L-function of Fig. M-1a does not describe the detailed electronic charge distribution $F(x)$ of the crystal, only its general periodicity. To obtain $F(x)$, we must take the unit-cell charge distribution function F_u and replicate it for all lattice points. This means taking the function F_u referred to the origin $(x = 0)$, duplicating it, shifting it to the lattice point $x = \delta$, and then repeating the same operations for all lattice points $(x = 2\delta, 3\delta, ...)$.

The mathematical tool for this procedure is the so-called 'convolution'. We can say that each shifted function F_u is the convolution of the original function F_u with a peak P centered at $x = n\delta$. The corresponding mathematical symbol is $F_u \otimes P$. To build the entire charge distribution $F(x)$, we must repeat the convolution for all lattice points. This is equivalent to the convolution of F_u with the function $L(x)$ of Fig. M-1a, which is the combination of all peak functions P:

$$F(x) = F_u \otimes L . \tag{M2}$$

We can now apply the already mentioned basic property of Fourier transforms: *the Fourier transform of the convolution of two functions equals the product of the Fourier transforms of the two functions*. The scattered wave $W(k)$ is the (inverse) Fourier transform of $F(x)$, and therefore equals the Fourier transforms of the L-function multiplied by the Fourier transform of F_u—which is by definition the *crystal structure factor $f(k)$*.

Thus, the scattered wave W is the product of the Fourier transform of L multiplied by the crystal structure factor $f(k)$. This justifies the experimental procedure outlined in the main text.

Specifically, the Fourier transform of the lattice-point function $L(x)$ gives the periodicity of the Bragg spot pattern. Each Bragg spot occurs at a specific direction corresponding to a given k-value. The wave in this direction is the Fourier transform of $F(x) = F_u \otimes L$ for the same k-value, and therefore it is proportional to the crystal structure factor $f(k)$. The square of the wave—which is determined by the square of $f(k)$—corresponds to the intensity of the Bragg spot.

In turn, by measuring the Bragg-spot intensity we can retrieve information on $f(k)$. Using a Fourier transform, we can then obtain information on the unit-cell charge distribution function $F_u(x)$ and determine the atomic structure of each cell in the crystal, i.e. of each molecule.

3.5.4. Powder diffraction: chemical and structural analysis

Many structural investigations are performed on samples which, although crystalline on a microscopic scale, are not ordered over macroscopic distances. These include, for example, polycrystalline specimens and fine ground powders. Crystallography in the corresponding 'powder diffraction' (Debye–Scherrer) mode provides very valuable information for such samples.

Figure 3.121 illustrates the effects on a diffraction pattern of the presence of many small randomly oriented crystals. As the number of microcrystals increases, the pattern evolves from single Bragg spots to circles. Diffraction data must then be taken not on spots but as plots of the diffracted intensity in a radial direction—see Fig. 3.122.

Such plots contain of course information about the Bragg spots that form the circles in the pattern. The information can be extracted by comparing the measured plots with reference plots for different compounds. The presence of a given chemical compound can thus be derived from the detection of the corresponding features in the plot. An exceedingly large amount of reference data is available from standard databanks.

Powder diffraction thus becomes an effective and versatile probe of the chemical composition. The potentially available information is not limited to the chemical composition. Powder diffraction can also yield excellent structural information on the crystal geometry, the crystal length parameters, the crystal structure factor, etc.

What are the advantages of using synchrotron sources for powder diffraction? First of all, their high brightness and collimation substantially improve the accuracy in measuring individual diffraction-pattern peaks—and the accuracy in extracting information. Furthermore, the wavelength-tunability of synchrotron sources can be exploited for special powder-diffraction techniques.

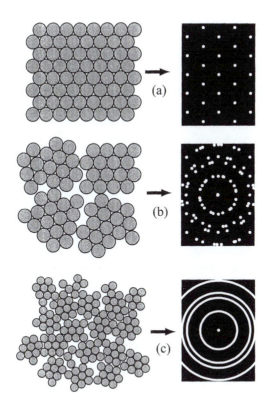

Fig. 3.121 Schematic explanation of powder diffraction: (a) a single crystal produces a diffraction pattern with Bragg spots; (b) and (c) as the single crystal is replaced by several randomly oriented small crystals, the pattern evolves into a series of circles.

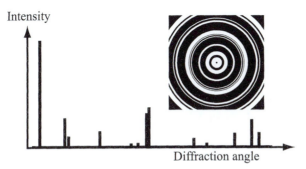

Fig. 3.122 Top right: a powder diffraction pattern. Bottom: radial intensity plot. From this plot, very valuable information can be extracted on the chemical composition and crystal structure of the specimen.

Specifically, the X-ray scattered intensity changes rapidly when the wavelength reaches an absorption edge of a given element. This can identify the contributions of that element to powder diffraction data.

3.5.4.1. A key problem; scattered intensity vs scattered waves

A more detailed analysis of powder diffraction leads us to *the* fundamental problem of X-ray structural studies. Such studies are based on obtaining structural information from the Fourier transform of the scattered wave. Unfortunately, this procedure is negatively affected by a crucial limitation.

A detector of X-rays—like a visible-light detector such as the human eye—does not reveal 'waves'. What is does detect is the *intensity I*, which is the power deposited by the wave per unit detector area. Basic electromagnetism shows that I is proportional to the *square* of the wave W:

$$I \text{ proportional to } |W|^2 . \tag{3.22}$$

This implies that intensity measurements do provide information about the wave W. However, the information about W is not complete because I is not equal to W. This significantly complicates the extraction of structural information from X-ray scattering and diffraction data. A simple example of this loss of information is discussed by Inset N.

The analysis of Inset N specifically shows that what is lost when we measure the intensity is the 'phase' of the wave. This loss negatively affects the structural analysis, causing the so-called 'phase problem'. As discussed later, the solution of such a problem makes synchrotron sources very desirable.

3.5.4.2. Patterson function

The 'phase problem' affects in particular X-ray powder diffraction. Ideally, if we could directly measure the scattered wave W, then the structural information could be extracted from W by performing a Fourier transform (eqn 3.19):

$$F = \text{Fourier transform of } W ;$$

this procedure would specifically yield the electron distribution function F, which, as we have seen, carries information both on the overall crystal structure and on the atomic positions inside each unit cell.

Unfortunately, the Fourier transform cannot be applied to the wave W, since what we measure is not W but the intensity I which is proportional to $|W|^2$. The Fourier transform of $|W|^2$ is called the 'Patterson function'.

The Patterson function does not directly correspond to the absolute positions of the scattering atoms as does the F-function. It corresponds instead to the *relative* positions of the atoms with respect to the other atoms (see Fig. 3.123 and Inset N). This information, although not directly equivalent to the determination of atomic positions, is still very valuable.

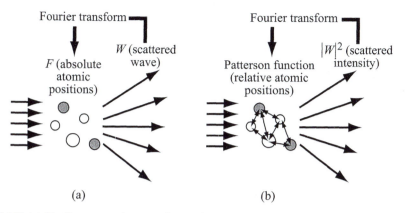

Fig. 3.123 (a) Ideally, a scattering experiment should measure the scattered wave W, and then retrieve (by Fourier transform) the electron charge distribution function F, which is related to the absolute atomic positions. (b) In reality, what we measure is the scattered intensity, whose Fourier transform yields the so-called Patterson function—which corresponds instead to the *relative* atomic positions.

Inset N: The phase problem and the Patterson function

The intensity, being linked to the square of the wave, carries less information than the wave itself. This fact can be illustrated by the simple case of Fig. K-1 (Inset K). The two point-like objects individually produce (eqn K-4) scattered waves proportional to the functions 1 and to $\exp(-ik\delta)$.

Figure N-1 shows plots of the real and imaginary parts of these two functions, that are equal to 1 and 0 for the (constant) function 1, and to $\cos(-k\delta)$ and $\sin(-k\delta)$ for the function $\exp(-ik\delta)$. We see that the differences between the two scattered waves are quite strong.

Consider now the square of the first scattered wave, which is by definition (Inset L) the square of the real part plus the square of the imaginary part. The result is proportional to $1^2 + 0^2 = 1$—shown by the dashed line of Fig. N-1a. As to the second wave, its square is proportional to $\cos^2(k\delta) + \sin^2(k\delta) = 1$. This is exactly the same result as for the first wave—see Fig. N-1b.

Thus, the two waves, although quite different from each other, have the same square and therefore correspond both to the same measured intensity. This illustrates how information is lost when the wave intensity is measured rather than the wave itself.

Why is this loss of information called the 'phase problem'? We must consider that a complex quantity like the wave W can be expressed in general as:

$$W = W_0 \exp(i\phi), \tag{N1}$$

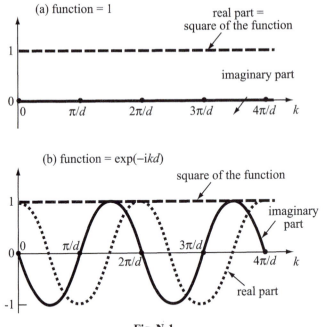

Fig. N-1

corresponding to the real part $W_0\cos(\phi)$ and to the imaginary part $W_0\sin(\phi)$. The angle ϕ is called the 'phase' of the complex number, and the term $\exp(i\phi)$ is called the 'phase factor', whereas W_0 is the 'amplitude' of the wave.

Equation N-1 can be easily applied to the two scattered waves discussed previously: the functions 1 and $\exp(-ik\delta)$. In the first case, $W_0 = 1$ and $\phi = 0$. In the second case, $W_0 = 1$ and $\phi = -k\delta$.

Consider now the square of Eq. N1. By definition:

$$|W|^2 = [W_0\cos(\phi) + iW_0\sin(\phi)]^2 = W_0^2[\cos^2(\phi)+\sin^2(\phi)] = W_0^2 . \quad (N2)$$

Therefore, by measuring the square of W we lose track of the phase ϕ. This is why the corresponding loss of information concerns the phase, and is called the 'phase problem'.

The 'phase problem' is present, in particular, when we consider the Patterson function. We can understand the properties of this function by using again the simple case of two atoms A and B—see Fig. N-2a. For simplicity, we assume that the electronic charge of each atom is concentrated at its center, so that the electronic charge distribution function $F(x)$ consists of two sharp peaks at $x = x_A$ and $x = x_B$ (see Fig. N-2b).

Note that the peaks do not have equal height. In fact, different atoms have different numbers of electrons and therefore scatter x-rays more or less effectively. We call the peak heights H_A and H_B.

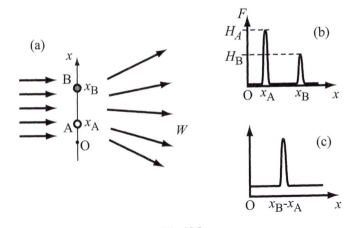

Fig. N-2

The scattered wave $W(k)$ is the (inverse) Fourier transform of the two peaks; according to the definition of eqn K6:

$W(k)$ proportional to $\int F(x)\exp(-ikx)dx$.

If the peaks are infinitely sharp, this simply gives:

$W(k)$ proportional to $H_A\exp(-ikx_A) + H_B\exp(-ikx_B)$.   (N3)

If we could directly measure this wave $W(k)$, then the position function $F(x)$ would be obtained by performing the Fourier transform of $W(k)$ (Equ. K-4):

$F(x)$ proportional to $\int W(k)\exp(ikx)dk$,

which is proportional to the integral:

$\int[H_A\exp(-ikx_A) + H_B\exp(-ikx_B)]dk$.

Because of the properties of sine and cosine products, this integral is zero except for $x = x_A$ and $x = x_B$—giving the two-peaked distribution function $F(x)$ of Fig. N-2b. Thus, the Fourier transform of W would identify the absolute atomic positions x_A and x_B.

The scattering experiment, however, does not measure $W(k)$ but the intensity, which is proportional to $|W|^2$. The (complex-number) square of eqn N3 gives:

$|W|^2$ proportional to $|H_A\exp(-ikx_A) + H_B\exp(-ikx_B)|^2$

$= |H_A\cos(-kx_A) + H_B\cos(-kx_B) + i[H_A\sin(-kx_A) + H_B\sin(-kx_B)]|^2$

$$= [H_A \cos(-kx_A) + H_B \cos(-kx_B)]^2 + [H_A \sin(-kx_A) + H_B \sin(-kx_B)]^2$$
$$= H_A{}^2[\cos^2(-kx_A) + \sin^2(-kx_A)] + H_B{}^2[\cos^2(-kx_B) + \sin^2(-kx_B)]$$
$$+ 2H_A H_B[\cos(-kx_A)\cos(-kx_B) + \sin(-kx_A)\sin(-kx_B)]$$
$$= H_A{}^2 + H_B{}^2 + 2H_A H_B\{\cos[-k(x_A - x_B)]\}\ .$$

The corresponding Fourier-transform integral (the Patterson function, Fig. N-2c) has a sharp peak for $x = x_A - x_B$, which is the relative position of one atom with respect to the other. Thus, the Patterson function reveals the relative distances between the atoms.

3.5.5. Synchrotron-based crystallography

Synchrotron sources play a central important role in X-ray crystallography. There are many aspects of this role: we focus our attention in particular on the very important subject of biological molecules.

3.5.5.1. Instrumentation: diffractometers

The instrumentation for synchrotron-based crystallography is becoming increasingly sophisticated. The general trend is towards automatic handling of many different tasks. The core instrument is the diffractometer, a device that automatically or semi-automatically records the diffracted intensity for different configurations—corresponding in particular to the directions of the Bragg spots.

The geometry changes are achieved by modifying several angles with accurate goniometers. Each adjustable angle corresponds to a circle; the standard nomenclature specifies for each diffractometer the number of circles. For example, a device with four different adjustable angles is called a 'four-circle diffractometer'. Figure 3.124 shows the scheme of a four-circle diffractometer: not only the Bragg angle ($2\theta_d$) can be changed, but also the three angles that define the sample position with respect to the X-ray beam.

In the specific case of Fig. 3.124, the diffractometer includes a small-area detector whose position can be changed. Other diffractometers use large-area devices such as CCDs that simultaneously detect the intensity of many different Bragg spots.

3.5.5.2. Practical problems in macromolecular crystallography

The general strategy for biomolecular crystallography was discussed in Section 2.5.3. We can summarize it as follows: (1) growth of an artificial crystal consisting of a large number of molecules; (2) measurements of the intensity of many different Bragg spots to obtain quantitative data on the crystal structure factor f, and then (3) extraction from such data of the electron charge distribution function F_u for an individual macromolecule (crystal cell). The function F_u reveals the atomic structure of the molecule.

The strategy requires crystals with suitable characteristics, including reasonably large size, resistance to radiation damage and good crystal quality. Growing such crystals is a task at the borderline between science and art—and a key problem in macromolecular crystallography.

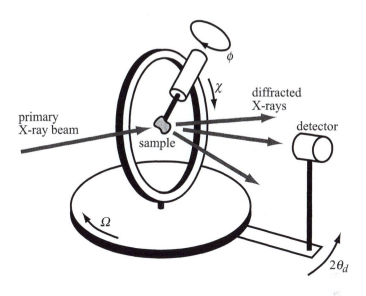

Fig. 3.124 Scheme of a four-circle diffractometer. The four goniometers are used to adjust the diffraction angle $2\theta_d$ and the sample position angles Ω, χ and ϕ. Note the horizontal scattering plane typical of laboratory-based diffractometers. For most synchrotron-based instruments the scattering plane is vertical to exploit the natural vertical collimation of the X-ray beam and avoid polarization-related problems.

Major breakthroughs such as cooling crystals to reduce damage substantially improved the situation. Focused X-ray beams, made possible by the excellent geometrical characteristics of synchrotron sources, open the door to the analysis of very small crystals. These improvements notwithstanding, the practical obstacles are very challenging and the number of unsolved molecular structures very large.

Important classes of molecules—notably, membrane proteins—cannot be crystallized or are very difficult to crystallize. In other cases, the long time required to grow a large crystal impacts on the total duration of a structure-solving procedure. This has important negative repercussions—for example, on the costs of developing new products in pharmaceutical research.

Another crucial problem is that the demand for biomolecular crystallography greatly exceeds the available synchrotron beamtime. Structural studies related to genomic programs and to drug design cause a rapid increase in demand. This produces a worldwide beamtime shortage and requires effective and flexible strategies.

Quite often, for example, the initially grown crystals do not possess the required quality level for advanced structural analysis. The first tests may thus give negative results resulting in a waste of valuable synchrotron beamtime. Therefore, such tests are often performed with laboratory X-ray sources, followed by advanced structure refinement at a synchrotron facility.

The flow of macromolecular structural data is becoming quite difficult to handle. This requires large-capacity data storage and high-flow transmission techniques. In the long term, the experts foresee automatically handled procedures with minimal human intervention. As a consequence, the robotization of the instrumentation is also an increasingly important technical issue.

3.5.5.3. Surface crystallography

This is a specialized crystallographic technique primarily used in chemistry and materials science but with potential impact on the life sciences. It exploits the advanced performances of synchrotron sources to investigate the atomic structure of crystal surfaces. In general, the surface structure is not equivalent to that of the bulk crystal, and can in fact be very different.

Consider (Fig. 3.125a) the situation in a bulk crystal: the atomic positions and therefore the entire crystal structure reflect an equilibrium state determined by the chemical bonds between atoms. This equilibrium state, however, is not the same for bulk atoms and for surface atoms, whose local chemical environments are quite different. As a consequence, the surface atomic positions are generally different from the equivalent bulk positions. This phenomenon is known as 'surface reconstruction' (see Fig. 3.125b) and can lead to a surface structure different from the bulk periodicity.

The experimental identification of surface structures is a very important issue since surface atoms play a key role in many chemical and materials science phenomena. These include, in particular, catalytic processes of crucial importance for industry.

The main problem in surface structural analysis is to separate the surface contributions to X-ray scattering from the much stronger bulk contributions. Special techniques with high surface sensitivity such as low-energy electron diffraction (LEED) or scanning tunnel microscopy (STM) automatically provide surface structural information. However, their accuracy seldom reaches the levels of X-ray crystallography.

The detection of surface-related signals in X-ray crystallography is essentially impossible with conventional X-ray sources. Surface crystallography becomes feasible with the high intensity and collimation of synchrotron sources. Figure 3.126 shows an example of this specialized but very important technique.

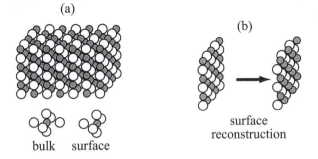

Fig. 3.125 The phenomenon of surface reconstruction: (a) the environment of a given atom is not the same in the bulk and at the surface; (b) as a consequence, the surface atoms can find a different equilibrium state corresponding to positions different from their bulk counterparts.

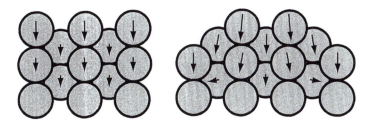

Fig. 3.126 A nice example of the applications of surface crystallography: this powerful technique was used in this case to detect fine displacements of the copper atoms at a copper surface, induced by the adsorption of cesium atoms. The figure illustrates two of the reconstructions analyzed by R. Schuster and I. K. Robinson, *Phys. Rev. Lett.* **76**, 1671 (1996).

3.5.6. Solving the phase problem

We mentioned that the phase problem (see Section 3.5.4 and Inset N) is the key obstacle in X-ray structural analysis. This is true in particular for macromolecular crystallography. This problem can be solved with different practical approaches, all of which belong to two broad classes.

The first type of approach (Fig. 3.127) is based on replacing some of the atoms in the unknown molecular structure with other atoms—preferably of heavy elements that scatter X-rays very effectively. The initial specimen is called the 'native' crystal, whereas the crystal with replacement atoms is called an 'isomorphous derivative'.

Crystallographic data must be taken both for the native crystal and for the isomorphic derivative. Under certain conditions the comparison of the measured Bragg spot intensities for the two crystals leads to a satisfactory solution of the phase problem—see the simple case of Inset O.

Specifically, it can be demonstrated that just one isomorphous derivative is not sufficient to remove all the uncertainties created by the phase problem. Measurements must then be performed on two or more different isomorphous derivatives. This may create serious difficulties: at least two additional crystals are required, whereas growing one type of crystal is already a challenging task.

These difficulties stimulated the development of the second class of solutions of the phase problem (see again Fig. 3.127). Such solutions are based on the so-called 'anomalous scattering' or 'anomalous diffraction' effect. These names are slightly misleading: there is nothing 'anomalous' in the underlying phenomenon. The X-ray scattering by a given atomic species changes with the X-ray wavelength. Such changes are typically small and slow. However, they become prominent and fast when the wavelength reaches one of the X-ray absorption thresholds of the atomic species. This is what is called 'anomalous scattering'.

This is shown for a copper sample in Fig. 3.128. The plots refer to the two parts of the so-called 'atomic structure factor' that describe X-ray absorption and X-ray scattering.

(a)

(b)

Fig. 3.127 Two solutions of the phase problem in macromolecular crystallography: in (a), measurements are performed both on the original (native) molecular crystal and on a new crystal (isomorphous derivative) in which some of the atoms are replaced by a different (heavy) atomic species. In (b), it is not necessary to grow an isomorphous derivative: experiments are made at different X-ray wavelengths, thus modifying the scattering response of specific atoms. This approach requires a wavelength-tunable synchrotron source.

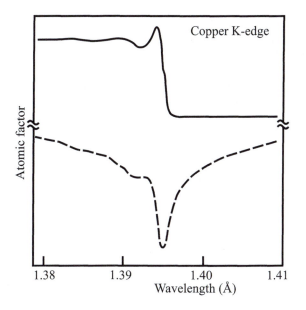

Fig. 3.128 Illustration of the rapid changes in the X-ray absorption and X-ray scattering by a copper specimen as the wavelength changes in the region of one of the copper absorption thresholds. Both absorption and scattering are described by the so-called 'atomic structure factor'. This factor includes a real part (dashed line) and an imaginary part (solid line). The imaginary part corresponds to the X-ray absorption whereas the real part describes X-ray scattering. Note that the changes in the real and imaginary parts, although qualitatively different, occur in the same narrow wavelength range and therefore appear correlated.

Note that the changes in absorption and scattering are correlated, i.e. they occur in the same wavelength range. This intriguing fact reflects a fundamental optical property valid for all systems: rapid changes in the X-ray absorption correspond to rapid changes in scattering. The correlation is one of the many manifestations of the 'Kramers–Krönig relations', briefly discussed in Inset F.

By taking scattering data at different wavelengths near an absorption edge, we can thus change the scattering response of the corresponding atomic species. This is somewhat similar to replacing the atoms of a different atomic species by growing an isomorphous derivative. The advantage, of course, is that changing the wavelength is much simpler than growing a new crystal.

The most sophisticated implementation of this strategy is the popular technique known as 'multi-wavelength anomalous diffraction' or MAD. Crystallographic MAD analysis requires taking data at multiple wavelengths. Therefore, it strictly requires a wavelength-tunable X-ray source. Because of the intensity problems, suitable sources are wigglers operating at medium-energy or high-energy synchrotrons.

MAD macromolecular crystallography yields the most accurate (refined) data on molecular structures—and is in very high demand. Figure 3.129 shows one of the many results produced by this approach.

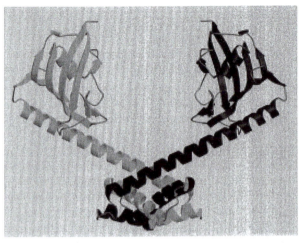

Fig. 3.129 Example of the applications of the MAD phasing technique for macromolecular crystallography: portion of the structure of the macrophage infectivity potentiator protein from *legionella pneumophila*. Data from: A. Riboldi-Tunnicliffe, B. König, S. Jessen, M. S. Weiss, J. Rahfeld, J. Hacker, G. Fischer and R. Hilgenfeld, *Nature Struct. Biol.* **8**, 779 (2001); copyright by *Nature*, reprinted by permission.

Inset O: Isomorphous derivatives and anomalous scattering

How can isomorphous replacement—or anomalous scattering—provide sufficient information to solve the phase problem? We cannot present here a detailed analysis, but we can use a very simple example to give a rough idea of the theoretical background.

We saw that the Fourier transform of the crystal structure factor f would give the electronic charge distribution—and therefore identify the structure of an unknown molecule. Measuring the crystal structure factor is thus the key objective of macromolecular crystallography. On the other hand, the crystal structure factor is a complex number with an amplitude and a phase factor (see eqn N1):

$$f = f_0 \exp(i\phi) . \tag{O1}$$

Unfortunately, the scattering experiments do not measure f but its square $|f|^2 = f_0^2$; they can reveal the amplitude f_0 of the structure factor f but not its phase ϕ.

Suppose, however, that a scattering experiment is performed for a native crystal and then repeated for an isomorphous derivative, corresponding to a different crystal structure factor:

$$f' = f_0' \exp(i\phi') . \tag{O2}$$

Once again, the experiment does not give f' but its amplitude f_0'. Assume that the replacing atoms contribute to f' with a structure factor $f'' = f_0'' \exp(i\phi'')$:

$$f' = f + f'' . \tag{O3}$$

Suppose that theory can calculate f''. For simplicity, we assume that in this particular case $f'' = 0$ and therefore $f'' = f_0''$, so that eqn O3 becomes

$$f_0' \exp(i\phi') = f_o \exp(i\phi) + f_0'' . \tag{O4}$$

By calculating the (complex-number) squares of the two sides of eqn O4 we obtain:

$$f_0'^2 = f_0^2[\cos^2(\phi) + \sin^2(\phi)] + f_0''^2 + 2f_0 f_0'' \cos(\phi) ,$$

and therefore

$$f_0'^2 = f_0^2 + f_0''^2 + 2f_0 f_0'' \cos(\phi) . \tag{O5}$$

All quantities in eqn O5 are known either by measurements or by calculations, except for $\cos(\phi)$. Therefore, this equation can be used to extract $\cos(\phi)$ and the phase angle ϕ. This information, however, is *almost* but not completely sufficient to derive the angle ϕ and therefore the complex crystal structure factor $f = f_0 \exp(i\phi)$. The remaining ambiguity affects the sign of the angle ϕ, since $\cos(\phi) = \cos(-\phi)$. This residual uncertainty can be removed by taking data on a second isomorphous derivative. Hence, we need two difficult-to-grow derivative crystals rather than just one.

Similar arguments can be used to explain the alternate strategy based on multi-wavelength measurements and anomalous scattering. We have already mentioned that the effects of anomalous scattering are, to some extent, similar to those of replacing atoms. Therefore, measurements at multiple wavelengths in a regime of anomalous scattering add information similar to those on isomorphous derivatives—without requiring the growth of additional crystals.

3.6. Microfabrication and other techniques

The rapid progress of synchrotron research keeps producing new applications. Before completing our review of important applications in chemistry, life sciences and medical research, we must briefly discuss a few such applications that are in most cases at a development stage.

3.6.1. Synchrotron X-ray lithography

Most synchrotron applications are analytical techniques: synchrotron X-rays are used as a probe of the chemical, atomic and electronic structure. Microfabrication, however, is an exception: its objective is not to characterize but to produce artificial microsystems.

The basis of synchrotron microfabrication is lithography for the microelectronics industry. This is the amazing technique that puts a huge number of tiny devices into each silicon wafer. Ultraviolet light is the fabrication tool, and the procedure is specifically called 'photolithography'.

Figure 3.130 illustrates the basic principles of photolithography with an idealized example. The objective is to reproduce on a silicon wafer a microscopic pattern (whichmight part of a device). The first step is to draw the pattern on a macroscopic scale and then to demagnify it while reproducing it in a mask.

The silicon surface—already coated by a passivating silicon dioxide layer and by a metal overlayer—is further covered with a layer of a photosensitive compound called 'photoresist'. The photoresist layer is then exposed to ultraviolet light passing through the mask. During this stage, the pattern can be further demagnified to its final dimensions by an optical system.

The ultraviolet light illuminates the portion of the photoresist corresponding to the pattern and changes its properties. In Fig. 3.130, we assume that the illuminated areas become resistant to chemical etching.

After exposure, chemical etching removes the non-illuminated photoresist and the corresponding portions of the metal overlayer. The remaining metallic coating reproduces the desired micropattern.

The many different photolitography techniques in microelectronics are variations of this basic approach. They are, therefore, all subject to similar problems and limitations.

A crucial problem is the minimum size of the pattern features. Excessively small mask features diffract the ultraviolet light and blur the reproduced pattern details. The minimum feature size is determined by the ultraviolet light wavelength that characterizes diffraction. For this reason, long-wavelength visible light was replaced long ago by shorter-wavelength ultraviolet light.

The obvious question is: Why not use even shorter wavelengths, replacing ultraviolet light with X-rays? The problem is that X-rays would lead to serious difficulties. For example, demagnification during illumination is difficult with X-rays because of the lack of suitable optical components—and same-size contact exposure must be used. Photoresist compounds for X-rays are quite difficult to obtain. The one that is most commonly used is polymethyl methacrylate (PMMA), whose sensitivity to exposure is limited. Furthermore, high-energy X-ray photons can produce a cascade of (secondary) free electrons in the photoresist that blur the pattern features.

Considering this last problem, the best wavelengths for X-ray lithography must be a compromise between limiting the diffraction effects and minimizing the secondary electron blurring. The optimal wavelength region is quite broad but it does not include conventional X-ray sources. Thus, X-ray lithography must primarily rely on synchrotron sources.

The above problems notwithstanding, X-ray lithography remains a promising technique for the microfabrication of electronic devices. As such, it could substantially impact chemistry, the life sciences and medical research. The corresponding applications range from routine to quite revolutionary approaches—for example, direct interfacing of microelectronics devices and living neuron systems.

In addition to the production of microelectronic devices, another use of X-ray lithography is particularly interesting for medical and biological applications. This is the so-called 'deep' lithography and in particular the technique called 'LIGA', an acronym explained in the next section.

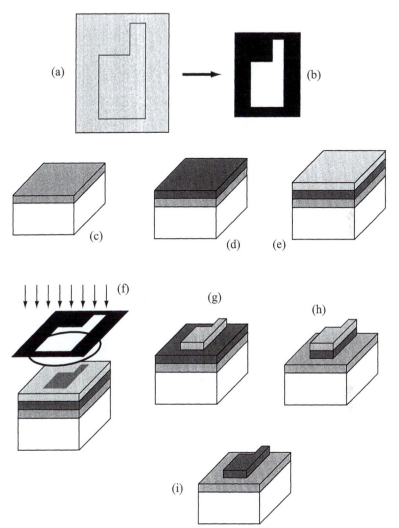

Fig. 3.130 Schematic and idealized picture of a photolithographic procedure: (a) the desired pattern is drawn on a macroscopic scale; (b) the drawing is reduced while producing a mask; the silicon wafer covered by silicon oxide (c) is coated with a metallic overlayer (d) and then with a photoresist layer (e). Ultraviolet light illuminates (f) the photoresist through the pattern mask with further demagnification by a lens; the illuminated parts of the photoresist change their chemical properties and become resistant to chemical etching (g) that eliminates the other photoresist parts. (h) The metal overlayer is chemically removed except where protected by the remaining photoresist; finally (i) the photoresist is entirely removed leaving the desired pattern.

3.6.1.1. Deep lithography: LIGA

The main difference between lithography for microelectronics and 'deep' lithography is that the latter works in three dimensions. A microelectronic device is basically a two-dimensional system. The features of each device layer are on the layer surface and the microfabrication techniques need not to work in depth.

On the other hand, 'deep' lithography does work in three dimensions leading to many new applications. Figure 3.131 shows two impressive examples of its capabilities. We see that deep lithography produces very small features, emulating microelectronics lithography. In addition, it maintains high lateral resolution and accuracy over a large depth.

This specific performance is characterized by the 'aspect ratio', a parameter combining the lateral resolution of the microscopic features and their height. A product like that of Fig. 3.131 is said to have a high aspect ratio.

What are the requirements for deep lithography? First, the X-ray beam must penetrate deep into the photoresist. This requires shorter wavelengths than the currently used microelectronics lithography techniques , i.e. X-rays instead of ultraviolet light. Second, the illuminating X-ray beam must travel in a well-defined direction and therefore be highly collimated. All such requirements are met by synchrotron-emitted X-rays. On the other hand, synchrotron sources for deep lithography need not be very sophisticated. Bending magnets with beam collimators are typically used rather than wigglers or undulators.

Fig. 3.131 Two examples of the impressive capabilities of deep X-ray lithography: miniaturized magnetic heads (top) and a Fresnel zone plate. Note the lateral accuracy of the fabricated parts (the horizontal scale bar corresponds in each picture to 0.01 mm), and also that the accuracy is maintained over a large depth. Images courtesy of Franco Cerrina, Center for NanoTechnology, University of Wisconsin-Madison.

The most important type of deep lithography is called 'LIGA', an acronym for the German words (*LIthographie, Galvanoformung und Abformung*) corresponding to 'lithography, electroplating and molding'. This is a three-step process schematically illustrated in Fig. 3.132. After years of refinement, the LIGA technology can now fabricate complicated and extremely accurate deep microstructures at a reasonable cost.

There exist several variations of LIGA and combinations of LIGA and other techniques. For example, the masks for LIGA can be fabricated by electron-beam lithography. This is a powerful technique but its costs are intrinsically high. Thus, it cannot be directly used for large-scale fabrication—but it can be usefully combined with LIGA.

Deep lithography is specifically used to fabricate optical components for synchrotron beamlines. Many of the devices discussed in Chapter 2 and Section 3.3 can in fact be produced with this technology—notably, focusing devices like the Fresnel zone plates. The most important applications of deep lithography are in micromechanics and specifically impact medical problems and the life sciences. For example, micropatterning is used to fabricate devices for cellular biology research.

In the long term, the most intriguing applications of deep lithography might be for controlled drug release. An integrated drug-release microdevice should include sensors to monitor the patient status (temperature, blood pressure, concentration of specific chemicals, etc.), a microprocessor to analyze the sensor data and reach suitable decisions and micromechanical devices to implement them.

All the corresponding fabrication technologies already exist except in part those for micromechanical devices. Microprocessors are extremely advanced and sensors have made impressive progress in recent years. Deep lithography is a promising way to reach a similar level of sophistication for micromechanical parts.

The affinities between microelectronics photolithography and deep lithography can facilitate the integration of different technologies in the same device. For example, the use of silicon for LIGA can lead to the integration of microelectronic controls in micromechanical parts.

Micromechanical fabrication could lead to an industrial revolution similar to microelectronics. This would affect not only drug delivery but also many other aspects of medical diagnostics and therapy, and of the life and chemical sciences in general.

3.6.2. Other applications of synchrotron light

Two other important classes of synchrotron-based techniques deserve to be mentioned before concluding our review. First, those in the general domain of photochemistry. Advanced synchrotron sources concentrate an unprecedented electromagnetic radiation power into small areas and volumes. The resulting photon-stimulated effects are very interesting and still largely unpredictable.

Reliable projections can to some extent be extrapolated from past results. Figure 3.133, for example, shows experimental results on the phenomenon called 'photostimulated desorption'.

In a photostimulated desorption process, a photon is absorbed and its energy is used to break the bonds that link adsorbed atoms and molecules to a solid surface. Very little is known about this mechanism in spite of its fundamental and practical importance.

High-intensity synchrotron light was required for the results of Fig. 3.133 which concern the desorption of *neutral* species rather than of *ionized* species. The detection of neutrals is indeed much more difficult than that of ions. Before using intense synchrotron light, the detection of neutrals in photostimulated desorption was impossible. Thus, only desorbed ions could be investigated.

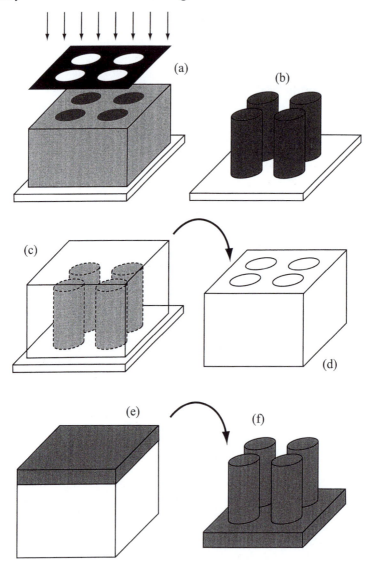

Fig. 3.132 Schematic explanation of the LIGA technology: (a) the photoresist (which coats a baseplate) is illuminated by X-rays through a mask; (b) after chemical development, only the illuminated areas of the photoresist are left ; (c) a metal mold is formed by electroplating around the remaining photoresist; (d) the mold is separated from the photoresist, and then (e) filled with plastic material; (f) the final part is extracted from the mold.

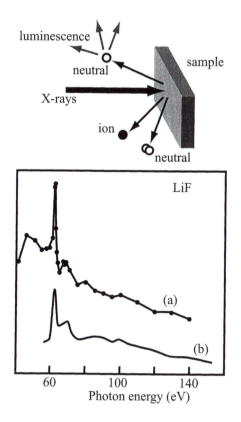

Fig. 3.133 Example of a photochemical (photostimulated desorption) experiment using synchrotron X-rays. Top: the photons break chemical bonds at a solid surface and stimulate the emission of neutral and ionized fragments. With suitably intense synchrotron light, the neutral particles can be detected by monitoring their luminescence as they travel away from the solid surface—and reveal very interesting properties. Bottom: (a) experimental data showing the luminescence intensity of Li atoms desorbed from a LiF surface as a function of the primary X-ray photon energy; curve (b) shows, for comparison, the LiF absorption coefficient in the same spectral region. Data from: N. H. Tolk, M. M. Traum, J. S. Kraus, T. R. Pian, N. G. Stoffel and G. Margaritondo, *Phys. Rev. Lett.* **49**, 812 (1982).

The first results on neutrals, made possible by synchrotron light, radically changed this point of view by showing that the emitted neutrals are much more abundant than the emitted ions. This is an example of how synchrotron sources can change our knowledge of photochemical phenomena leading to new discoveries and new applications. As a consequence, synchrotron-based studies related to photochemistry are a rapidly expanding domain.

We should note, in particular, time-coincidence experiments in which several correlated particles are detected. This approach substantially increases the amount of information on the corresponding photon-stimulated process—see Fig. 3.134.

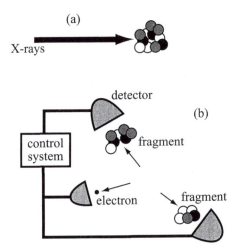

Fig. 3.134 (a) A hypothetical photochemistry effect that produces fragments and other electrons after the absorption of an X-ray photon. (b) The coordinated detection of different particles yields valuable information on the nature and dynamics of the phenomenon.

The low signal level is a severe problem in time-coincidence experiments. The high intensity of synchrotron radiation is in most cases an essential solution to this problem.

Finally, we should mention that synchrotron X-rays are used in medical research not only as analytical/diagnostic probes or fabrication tools, but also as therapeutic instruments. Interesting studies are being performed on the interaction mechanisms of intense X-rays with living organisms—to explore for example the selective destruction of cancer cells. The results, although still limited, are quite interesting and promising. Practical future applications could include not only the destruction of cancer cells by irradiation but also the direct use of X-ray beams for microsurgery.

4. Free electron lasers (FELs)

Synchrotron sources emit radiation with properties similar to lasers—for example, the undulators produce collimated, monochromatic and (partly) coherent light. On the other hand, the working principles of synchrotrons are different from those of lasers.

There exist, however, a new class of laser sources that are related to synchrotrons, namely the free-electron lasers or FELs. FELs are true lasers but their technology is strongly linked to that of non-lasing synchrotrons.

FELs are becoming particularly important for chemistry, biology and the life sciences. Until now, their emission was primarily confined to wavelengths in the infrared domain. However, some new FELs operate in the ultraviolet and soft X-ray domains. Devices emitting even shorter-wavelength X-rays may soon become reality.

A discussion of synchrotron research would not be complete without the basic notions on FELs and on their applications. Presenting them is the objective of this chapter. We begin with a simple description of lasers in general and then of FELs, followed by a discussion of specific FEL applications that are particularly relevant to this book.

4.1. Elementary description of lasers and FELs

Lasers have become quite common and many people are familiar with their basic features: well-defined color (i.e. well-defined wavelength), extremely low angular divergence and high intensity. Those interested in holograms may also know something about coherence.

However, only a few people are familiar with the working mechanism of lasers. Therefore, before discussing FELs we must briefly illustrate how a normal laser works. The underlying phenomenon is the so-called 'optical amplification', schematically illustrated in Fig. 4.1. This figure must be considered in relation to Fig. 3.3 and the discussion of the absorption of electromagnetic waves (eqn 3.1) in Chapter 3.

In that discussion we implicitly assumed that absorption and transmission are the only possible results of the interaction between a photon beam and the surrounding medium. This, however, is not always true. In some cases, the beam is neither absorbed nor simply transmitted but increases its intensity while traveling through the medium. When this occurs, there is optical amplification: energy is given by the medium to the wave rather than the other way around.

By exploiting optical amplification, we can transform a low-intensity wave into a high-intensity laser beam. This is realized in practice by using an 'optical cavity' formed by two mirrors (see Fig. 4.2). One is a fully reflecting mirror, whereas the other is 'semi-transparent', i.e. it transmits part of the wave and reflects the rest.

What is the role of the optical cavity? Consider a wave traveling through the optically amplifying medium. The amplification takes place along the entire wave path inside the medium and thus it increases with the path length.

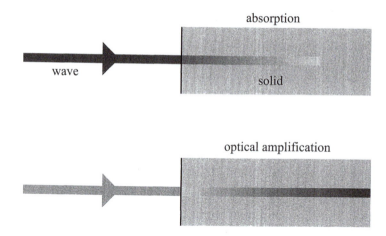

Fig. 4.1 Top: normally, when a wave reaches a solid (or another medium) it is either transmitted or—as in this case—absorbed. Bottom: under certain circumstances the wave intensity is increased by 'optical amplification'.

The optical cavity increases the effective path length by forcing the wave to travel several times through the medium. The wave loses part of its intensity each time it reaches the semi-transparent mirror. The rest of the intensity is reflected, then amplified as the wave travels along the length H of the medium, fully reflected by the other mirror and further amplified before reaching again the semi-transparent mirror. Thus, between two subsequent intensity losses the wave is amplified along a total path $2H$.

The production of intense laser light with this approach requires the amplification to offset the losses. The wave loses intensity whenever it reaches the semi-transparent mirror which thus becomes the source of laser light. Losses are also caused, for example, by impurities and defects in the amplifying medium. The basic lasing condition is that the optical amplification in the cavity must offset the combination of all losses.

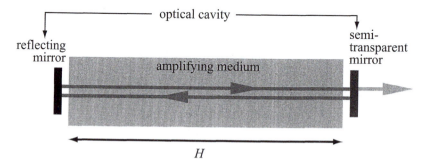

Fig. 4.2 An optical cavity—formed by a reflecting mirror and by a semi-transparent mirror—enhances optical amplification by increasing the effective path along which the amplification takes place.

The core component of a laser is the amplifying system. There exist several types of amplifying systems corresponding to different types of lasers: dye lasers, gas lasers, solid-state lasers, laser diodes, etc. For all lasers, the mechanism leading to optical amplification is 'stimulated emission'.

To understand its nature, consider first non-stimulated or 'spontaneous' emission (Section 3.1.1 and in particular Fig. 3.12): a system in an excited state decays to a lower energy emitting a photon. The decay and the photon emission occur spontaneously without any external stimulation. On the other hand, the decay and the photon emission can also be stimulated by the arrival of a pre-existing wave—see Fig. 4.3a. In this case, 'stimulated' emission reinforces the intensity of the wave, causing optical amplification.

Why does stimulated emission occur? Consider the basic requirement for the emission of electromagnetic waves (Section 1.1.2.2): the accelerated motion of electric charges. Specifically, the emission of a sinusoidal wave with a given frequency requires electric charges oscillating at the same frequency.

This oscillating-charge picture is the classical physics description of wave emission. How can we reconcile it with the quantum picture of jumps between energy levels? Quantum physics provides the answer: as indicated in Fig. 4.3b, during a jump between two levels an electron becomes equivalent to a classic oscillating charge.

The equivalent electron oscillation can start spontaneously, causing spontaneous emission. On the other hand, it can also be triggered by the arrival of a wave with the same frequency. This produces stimulated emission.

Our analysis reveals several important laser properties. First, to produce strong optical amplification the amplifying medium must have many excited electrons. This requires injecting energy into the medium and can be implemented in several different ways. In a gas laser, for example, the energy can be injected by an electric discharge.

Second, the jump ΔE between the energy levels determines the frequency of the amplified wave. In fact, the stimulated oscillation leads to the emission of photons of energy ΔE, and (eqn 1.5) the corresponding oscillating frequency is $\nu = \Delta E/h$.

Third, stimulated emission occurs in the same direction as the stimulating wave—whereas spontaneous emission has no preferential direction. The optical cavity reinforces this directional character: the laser output is highly collimated and therefore very intense.

Finally, the wave produced by stimulated emission is closely related to the stimulating wave. In particular, the oscillations of the two waves are in phase with each other. This—combined with the monochromatic character, the small source size and the low angular divergence—implies a high level of coherence.

Assume now that all the lasing conditions exist; how does the laser action actually start? The answer is that spontaneous emission can produce an initial wave. Then, this wave is amplified by stimulated emission enhanced by the optical cavity. This leads to the emission of intense laser light from the semi-transparent mirror.

4.1.1. Free electrons as optical amplifiers

Lasers are no longer exotic devices for specialized laboratories: they are used in CD players, laser pointers and many other consumer products. Such devices use a variety of amplifying systems including solids, gases, liquids and artificial nanostructures.

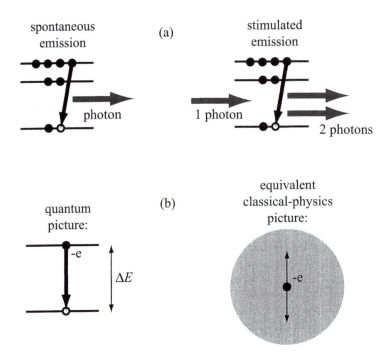

Fig. 4.3 (a) Comparison between spontaneous and stimulated emission. In the second case, the jump between energy levels that causes the emission of a photon is stimulated by a pre-existing wave whose intensity is thus amplified. (b) The quantum picture (left) for the jump of an electron between energy levels is equivalent in classical physics (right) to an oscillating electron charge. The result in both pictures is the emission of waves.

In all amplifying systems the active elements are electrons. The medium—liquid, solid or gas—constitutes to some extent only a support for the optically amplifying electrons. However, the medium also causes losses and other problems for the laser operation—for example by limiting the output power.

Could we then *eliminate the support* and build an optical amplifier made only of active electrons? The answer is positive: this is the basic principle of FELs.

Figure 4.4a shows an FEL facility, whereas Fig. 4.4b schematically shows its main components: an electron accelerator, a wiggler (planar or elliptical) and an optical cavity. Except for the optical cavity, such components are quite similar to a standard synchrotron source.

An FEL, in fact, works to some extent as a standard wiggler source: the electrons undulate in the transverse direction and spontaneously emit electromagnetic waves in the forward direction. In addition to this spontaneous emission, we must consider stimulated emission. The electrons (Fig. 4.5) travel along the wiggler axis together with the waves they have emitted. Such waves interact with the electrons producing stimulated emission that can lead to optical gain.

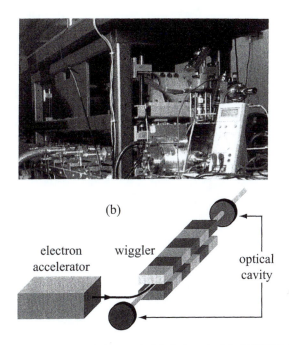

Fig. 4.4 (a) The wiggler at the core of the Vanderbilt University Mark III FEL (picture courtesy of David Piston). (b) The main components of an FEL: electron accelerator, wiggler and optical cavity.

Under appropriate conditions, the optical gain offsets the losses and the device works as a laser. Since the electrons are not bound to atoms or molecules but 'free', such a laser is immune to many of the limiting conditions of standard lasers. For example, the emitted wavelength is not fixed but can be modified according to the undulator emission law (eqn 1.8):

$$\lambda_L \approx \frac{L}{2\gamma^2}\left(1+\frac{K^2}{2}\right).$$

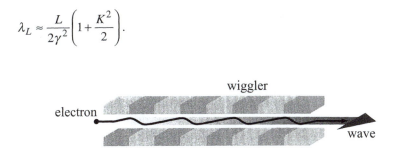

Fig. 4.5 Each electron in an electron bunch passing through the magnet array of an FEL travels together with the waves previously emitted by the bunch. The wave–electron interaction leads to stimulated emission and can produce optical gain.

This can done by changing the *K*-parameter (i.e. the magnetic field strength) or γ (i.e. the electron energy).

Several factors can help to achieve the required optical gain and transform a wiggler or an undulator into an FEL. As in standard lasers, the gain is enhanced by the optical cavity. Additional conditions increasing the optical gain in FELs are:

1. A well-defined electron beam energy. In a synchrotron source, different circulating electrons have slightly different energies. If this 'energy spread' is too large, then the optical gain can be jeopardized.
2. A small transverse cross section of the electron beam.
3. Small relative angular deviations of the trajectories of the individual electrons.
4. A large number of free electrons, i.e. a sufficiently high intensity of the electron beam current.

It can be also demonstrated that the optical gain *increases with the wavelength*. This is a very important point: short-wavelength FELs for X-rays are much more difficult to obtain than long-wavelength FELs. In fact, the majority of the FELs operate at long wavelengths in the infrared region.

The requirement of a high electron beam current leads to an additional important point. Given a certain *average* number of electrons passing through the wiggler, it is better to 'bunch' them together rather than spreading them apart. A shorter bunch corresponds in fact to a higher *peak* current.

We have already seen (Fig. 1.8) that a storage ring with a radiofrequency cavity squeezes the circulating electrons into bunches. The same conclusion is valid for other types of electron accelerators used for FELs.

In addition to the accelerator-caused bunching, another bunching phenomenon facilitates the FEL operation. The wave–electron–wiggler interaction tends to arrange each electron bunch into a series of 'microbunches' (see Fig. 4.6 and Inset P). This enhances the peak current of each microbunch and the optical gain.

4.1.1.1. Types of FELs

The general FEL scheme of Fig. 4.4b can be realized as a practical device using several different approaches and configurations. The wiggler, for example, can be either planar or elliptical.

electron
bunches

Fig. 4.6 The microbunching effect: owing to the interaction between the electrons, the magnet array and the wave, each electron bunch is reshaped as a series of shorter microbunches. This increases the peak current and enhances the optical gain.

The electron accelerator can be a storage ring as in standard synchrotron facilities. However, FELs are most often based on other types of accelerators such as electrostatic devices or linear accelerators (LINACs).

We have seen that given the conditions for optical gain the laser mechanism is triggered by an initial wave produced by spontaneous emission. There is an alternate approach: the initial triggering wave can also be externally produced—for example by a standard laser—and sent into the optical amplification system. The use of such a 'seed wave' facilitates the control of the FEL emission characteristics.

4.1.1.2. FEL properties

The light emitted by an FEL shares with synchrotron sources important characteristics, such as small angular divergence, small source size, monochromaticity and high coherence. In addition, the FEL can reach extremely high brightness levels.

Note, however, that this increase does not directly concern the *average* brightness. What increases is the *peak* brightness of very short pulses corresponding to the bunched electron beam—see Fig. 4.7.

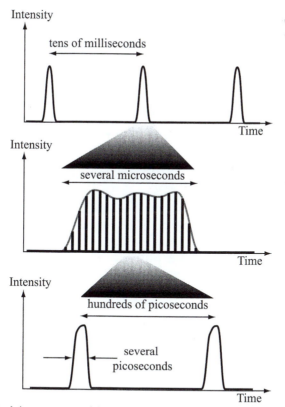

Fig. 4.7 The typical time structure of the emission of an FEL (the reference times in this case are from the Vanderbilt University infrared FEL). The structure consists of macropulses; each macropulse is formed by a series of micropulses.

A comparison between FELs and standard sources is particularly interesting in the case of infrared emission. The FELs produce short pulses of very high peak brightness (and good average brightness). Furthermore, such pulses are spectrally tunable: relatively simple changes in the FEL parameters can modify the output wavelength over a wide range.

Inset P: FEL mechanism

The complete FEL theory is quite complicated and beyond the scope of this book. Nevertheless, simple qualitative considerations can help us to better understand its foundations.

A standard wiggler or undulator emitter is basically an energy converter: it takes part of the energy of the passing electrons and converts it into electromagnetic waves. The energy conversion is produced by the magnet array that forces the electrons to undulate in the transverse direction. The undulation is caused by the Lorentz force applied by the magnets to the electron charges.

Note that the Lorentz force is perpendicular both to the magnetic field and to the electron velocity. A force perpendicular to the velocity cannot produce any work. Therefore, it cannot directly give or take energy to or from the electron. In other words, the energy exchange in a wiggler does not directly involve the magnet array: it occurs between the electrons and the wave with the magnet array acting as the mediator.

A similar conclusion is valid for stimulated emission in an FEL. Imagine (Fig. P-1a) an electron and a wave traveling along the same perfectly straight path. The magnetic field of the wave would act on the electron with a Lorentz force that cannot produce any work. The electric field of the wave would cause a force on the electrons in the transverse direction. As long as the electron velocity remains longitudinal, this transverse force cannot produce any work. In summary, there would be no energy exchange between the electrons and the wave.

This, however, would not be true if the electron velocity had a transverse component. The transverse force of the wave's electric field would then produce work, causing the transfer of energy between the wave and the electrons.

How can we obtain a transverse component of the electron velocity? We must add to the picture the FEL wiggler magnet array (Fig. P-1b). Its Lorentz force cannot produce any work. However, it causes transverse undulations of the electron trajectory with a transverse velocity component, and enables the wave's electric field to produce work. In summary, the FEL magnet array does not directly participate in the energy exchange, but mediates the electrons–wave energy exchange that leads to stimulated emission and to optical gain.

We must now consider another important aspect of the FEL mechanism: microbunching. The effectiveness of stimulated emission in producing optical gain depends on the peak electron beam current, i.e., on the simultaneous and correlated emission of many electrons. In turn, the peak current is enhanced by electron microbunching. The microbunching phenomenon can be understood with simple qualitative arguments.

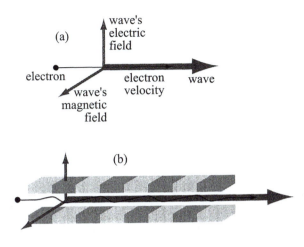

Fig. P-1

Consider (Fig. P-2) at a certain time electrons interacting both with the magnet array and with the electromagnetic wave. The electrons undulate in the transverse direction because of the action of the magnet array—thus their velocities have transverse components. The magnetic field of the wave interacts with these transverse components causing a longitudinal Lorentz force.

The direction of the Lorentz force changes with the direction of the wave's magnetic field. As shown in Fig. P-2, this force tends to squeeze the electrons into periodic microbunches with a period equal to the wavelength. This is the cause of the microbunching mechanism.

A more careful analysis, however, reveals two problems in this qualitative approach. First, the microbunching would apparently pile up electrons at the zero-field points of the wave. This would eliminate the wave–electron interaction required for stimulated emission.

Second, because of the undulation caused by the magnet array the transverse velocity periodically changes its direction. We could then conclude that the microbunching forces of Fig. P-2 would be reversed in the next half-undulation, no longer producing the same microbunching.

Both of these problems disappear if we consider an additional point in the electron–wave interaction. Whereas the wave travels exactly at the speed of light, the electron speed is close to c but slightly smaller. Thus, the positions of the electron microbunches shift continuously with respect to the wave. The microbunches, therefore, cannot travel together with the zero-field points of the wave—and the first problem disappears.

The wave–electron velocity difference causing the continuous shift between each microbunch and the wave is $(c - u)$. During the time required for the wave to travel along one period L of the magnet array:

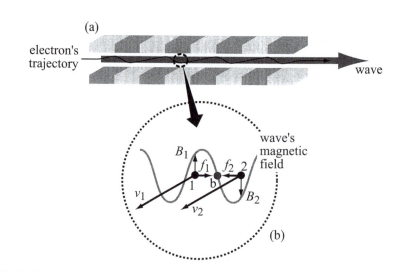

Fig. P-2 Simplified explanation of the microbunching mechanism: (a) the undulations induced by the periodic magnet array create a transverse component of the electron's velocity perpendicular to the wave's magnetic field; (b) this produces a Lorentz force that tends to bunch the electrons together. The enlarged picture shows two electrons (1 and 2) at places corresponding to different directions of the wave field; v_1 and v_2 are the transverse components of the electron's velocity; B_1 and B_2 show the wave's magnetic field acting on the two electrons; f_1 and f_2 are the corresponding Lorentz forces. Note that these forces tend to bunch the electrons at one single position (e.g. point b) for each wave oscillation — producing a periodic series of microbunches with period equal to the wavelength.

$$\Delta t = L/c \,,$$

the wave–electron shift is

$$\text{shift} = (c - u)\Delta t = (c - u)(L/c) \,.$$

This expression can be manipulated in the following way:

$$\text{shift} = (c - u)(L/c) = (1 - u/c)L = \frac{1 + u/c}{1 - u/c}(1 - u/c)L = \frac{1 - u^2/c^2}{1 + u/c}L$$

$$= \frac{L}{\gamma^2(1 + u/c)} \,,$$

which, since $u/c \approx 1$, becomes:

$$\text{shift} \approx L/(2\gamma^2) = \lambda_L \,. \tag{N1}$$

This is an interesting result: along one period of the magnet array the shift between the wave and an electron microbunch equals one wavelength λ_L (see eqn 1.4). Therefore (see Fig. P-3), during each half-undulation $L/2$ the shift is $\lambda_L/2$. Thus, the change in direction of the transverse electron velocity is compensated by the change in direction of the wave field, and the second of the above problems also disappears.

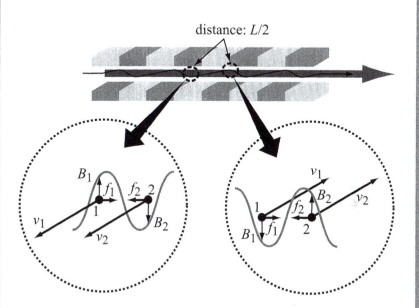

Fig. P-3 Consequences of the shift between the wave and the electron described by eqn N1. During one-half undulation $(L/2)$ the shift is one-half wavelength. Thus, not only the transverse velocity but also the wave fields change direction, leaving the force directions unchanged.

Our simplified analysis, however, is still affected by a problem: it could give the impression that stimulated emission always produces high optical gain and FEL action. This is not true, since the actual gain depends on a number of additional factors, including those mentioned in the text (high current, good electron beam geometry, etc.) and others. The actual FEL operation is only possible if all such factors are reasonably good.

4.2. Applications of infrared FELs

The high FEL optical gain at long wavelengths facilitates, as we have seen, the operation of infrared FELs with respect to ultraviolet and X-ray FELs. A large worldwide network already exists of infrared FEL facilities, many of which are open to users. The corresponding wavelengths range from tens of microns to a fraction of one micron. The user community is expanding and very active in different domains, including in particular the areas relevant to this book.

4.2.1. FEL surgery

Conventional lasers are already widely used in specialized surgical procedures. Infrared FEL beams were proposed several years ago for specific surgical applications. The first practical tests were very successful. What advantages can FELs offer with respect to standard surgical tools or other lasers? In addition to the high intensity level, the main strength is the possibility of selecting the wavelength over a broad spectral range. This tunability can be exploited to optimize the wavelength for each specific surgical application.

The wavelength optimization must be a compromise between effectiveness in cutting the target tissue and the need to avoid collateral damage to surrounding tissue. The initial tests to optimize the wavelength were at $\lambda \approx 3$ microns, corresponding to water absorption. This wavelength, however, causes substantial collateral damage.

Subsequently, the tests were focused on a longer wavelength, around 6.45 microns, corresponding to absorption both by water and by many protein molecules. The practical tests were quite encouraging—see Fig. 4.8. Other interesting wavelengths were identified later, notably $\lambda \approx 7.5$ and 7.7 microns for high-precision bone surgery.

The first human surgery with an FEL was performed at the Vanderbilt University Free Electron Laser Center in January 2000. The successful operation removed part of a brain tumor. Subsequently, other surgical procedures were tested at the same center, including eye surgery (optic nerve sheath fenestration before eye enucleation).

Surgery requires an FEL with very high stability and reliability and a sophisticated FEL beam delivery system. Conventional optic fibers are affected by problems because of the high power of the delivered beam.

A suitable solution is provided by waveguides formed by hollow glass tubes with a highly reflective coating on the interior. Each waveguide is coupled to a lens that delivers the focused beam into a spot size of the order of a few hundred microns.

Fig. 4.8 Direct comparison of the damage to surrounding tissue caused by surgical tests at two different FEL wavelengths. The damage was quite limited for $\lambda \approx 6.45$ microns (left), whereas (right) similar tests with $\lambda \approx 3$ microns produced substantial damage. Results from: G. Edwards, R. Logan, M. Copeland, L. Reinisch, J. Davidson, B. Johnson, R. Maciunas, M. Mendenhall, R. Ossoff, J. Tribble, J. Werkhaven and D. Oday, *Nature* **371**, 416 (1994); copyright by *Nature*, reprinted by permission.

Figure 4.9 shows pictures from early surgery tests at the Vanderbilt FEL. In order to support FEL surgery, the Vanderbilt center is equipped with a full clinical facility inside the FEL building. This radically changes the overall facility with respect to synchrotrons or FELs that are entirely dedicated to research.

What could be the future of FEL surgery? This is a difficult question to answer. FELs are excellent tools since they deliver very high power over a broad spectral range from which the optimal wavelength for each surgery can be selected. The Vanderbilt FEL, for example, yields over its broad range a peak power of more than 10 MW and an average power exceeding 10 W (with a light pulse length of less than one billionth of a second).

However, suitable high-power conventional lasers could be developed for each of the optimized wavelengths. Their use would be easier and more flexible than FELs. We can then foresee that tunable FELs might be primarily used to identify the best wavelength for each surgical procedure and guide the development of suitable conventional lasers for that wavelength.

On the other hand, FELs are also likely to continue to be used for specialized surgical procedures. For example, they appear the best solution for operations requiring very high power or several different wavelengths.

4.2.2. FEL infrared spectroscopy

The conceptual background of infrared spectroscopy using FELs is quite similar to that for synchrotron sources. Its main elements were discussed in Section 3.2.4.

The reasons in favor of using synchrotrons as infrared sources are also valid for FELs. In addition, the high FEL peak power opens new applications in non-linear spectroscopy.

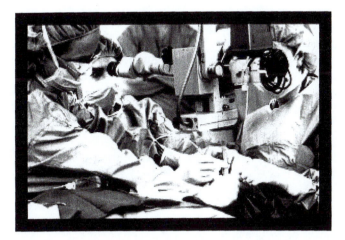

Fig. 4.9 A pioneering eye surgery test at the Vanderbilt University Free Electron Laser: Dr Karen Joost (second from left) operates with the FEL, assisted by Dr Louise Mawn (left). Photo by Dana Johnson; see: David Salisbury, Exploration (Vanderbilt University, Oct. 9, 2001) [http://www.vanderbilt.edu/exploration/news/news_fel_surgery_nsv.htm].

This term identifies phenomena whose magnitude increases very rapidly with the laser power. More specifically, the magnitude of most laser-power-triggered phenomena is proportional to the power. Non-linear phenomena increase instead more rapidly than a proportionality law. Therefore, they take advantage of the concentration of the FEL output into short pulses with high peak power.

Among the many non-linear effects, we note two-photon absorption. In this phenomenon, the jump of an electron between quantum energy levels is due to the simultaneous absorption of two photons instead of one. Two-photon absorption spectroscopy delivers information that is largely complementary to standard absorption spectroscopy.

Finally, the FEL time structure (Fig. 4.7) is exploited in time-resolved studies. For example, vibrational spectroscopy can analyze phase transitions in real time.

4.2.3. FEL microscopy

The high power of infrared FELs is very useful in experiments at the microscopic scale. The best performing instrument for FEL microscopy and spectromicroscopy is the so-called 'scanning near-field optical microscope' (SNOM or NSOM). Originally developed for visible light, the SNOM was then successfully extended to infrared FEL light.

The main objective of SNOM is to overcome the 'diffraction limit' that affects classical microscopy. We have already mentioned this limit while discussing lateral coherence (Section 1.6.1). In classical microscopy, the diffraction limit describes the fact that when the imaged details are too small diffraction blurs them.

Diffraction depends on the wavelength λ. Practically speaking, the diffraction blurring limits the lateral resolution of a standard magnifying system to a value determined by λ—see Fig. 4.10a.

The diffraction limit is a fundamental optical property and cannot be overcome by technical improvements within classical microscopy. The limit can be shifted to lower values by using shorter wavelengths. This led to electron microscopy, since the typical electron wavelengths used in microscopy are shorter than visible light. Similarly, the diffraction limit becomes less stringent when X-rays are used instead of visible light.

The diffraction limit affects classical microscopes but *not* all microscopes. Specifically, it does not apply to instruments working in the 'far-field' geometry.

The notion of far-field geometry can be understood with an example from medicine: the stethoscope (Fig. 4.10b). When a doctor uses a stethoscope, he/she easily finds the heart position with accuracy better than, say, one centimeter. This is well beyond the diffraction limit: the frequency of the sound waves of the heart (due to the heartbeat) is of the order of 1 pulse per second (1 Hz). The corresponding wavelength is the speed of sound (300 m/s) divided by the frequency, which gives $\lambda \approx 300$ m.

Therefore, the lateral resolution of a stethoscope is more than four orders of magnitude better than the diffraction limit! How is that possible? The answer is that the detection instrument (the stethoscope) and the wave source (the heart) are small with respect to λ and closer than λ to each other.

Similarly, the diffraction limit for light can be overcome by placing a small-area detector very close to a small source. This geometry is illustrated in Fig. 4.11 together with two practical SNOM approaches.

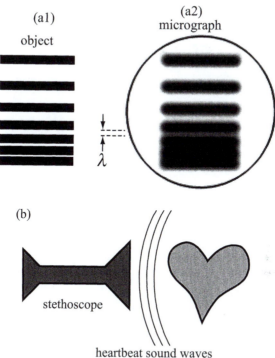

Fig. 4.10 The diffraction limit and one of its apparent violations. An object (a1), when imaged by a standard microscope (a2), is blurred by diffraction. The resolution is approximately limited to values larger than the wavelength. However, a stethoscope (b) localizes the heart with lateral accuracy much better than the sound wavelength. The reason is that the stethoscope works as a near-field microscope.

Near-field microscopy with light is schematically explained by Fig. 4.12. We assume that the source is an optic fiber with a narrow end as in Fig. 4.11b.

If the fiber tip is very small, the light cannot be transmitted through it. Even so, a small amount of light can still penetrate into the region immediately beyond the tip, without further propagation.

This 'evanescent wave' does not spread in the lateral directions and therefore produces a very localized illumination. When microimages are taken, the resolution is determined by the localization of the illumination rather than by diffraction.

A rigorous treatment of the SNOM approach is somewhat more complicated—see Inset Q. However, the basic ingredients are still similar to the stethoscope: a small source size (and/or a small detector size) and close proximity between source and imaged specimen.

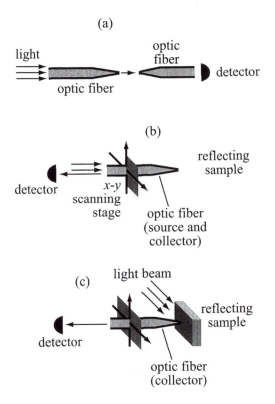

Fig. 4.11 (a) A hypothetical SNOM geometry equivalent to the stethoscope of Fig. 4.10b; (b) a practical SNOM geometry: the small source is an optic fiber with a narrow tip very close to the sample; the reflected light is collected by the same fiber; (c) an alternate SNOM geometry: the sample is illuminated by a light beam independent of the fiber.

The use of SNOM with an FEL requires solving several practical problems. Very high-quality optic fibers with narrow tips are needed that must resist high power. Other problems are similar to those of SNOM with conventional sources. In both cases, for example, SNOM micrographs must always be taken in conjunction with 'topographic' micrographs of the same area. This is needed to distinguish true optical features in the SNOM micrographs from surface topography artifacts.

Figure 4.13 shows the picture of a SNOM module that was developed for an infrared FEL. Experiments with modules like this produce two types of results. First, SNOM micrographs—see for example Fig. 4.14. A line scan (Fig. 4.14b) reveals that the lateral resolution is well below the wavelength and therefore beyond the diffraction limit.

To obtain the second type of results, SNOM pictures are taken with monochromatic (one-wavelength) infrared light. This emphasizes the absorption by specific vibrational modes. Such modes correspond to chemical bonds involving specific chemical elements.

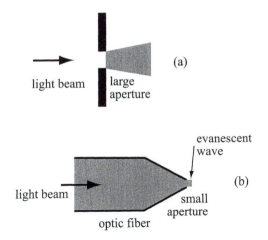

Fig. 4.12 Schematic explanation of SNOM: (a) when a large aperture is illuminated with a divergent beam, the light propagates beyond the aperture with a large lateral spread; (b) if the aperture is very small—such as the narrowed tip of an optic fiber—the light can only penetrate into a region very close to the aperture, and the lateral spread is very limited. Thus, the illumination is extremely localized.

Thus, SNOM images of this type carry chemical information similar to photoemission spectromicroscopy. For example, they reveal the lateral distribution of specific elements in specific oxidation states. Figure 4.15 shows an example of this 'spectroscopic SNOM' approach.

FEL-based SNOM finds interesting applications in the life sciences. This is particularly true for microchemical analysis with spectroscopic SNOM.

Fig. 4.13 Experimental multipurpose SNOM module used with an FEL source; see: A. Cricenti, R. Generosi, C. Barchesi, M. Luce and M. Rinaldi, *Rev. Sci. Instrum.* **69**, 3240 (1998). Photo courtesy of A. Cricenti.

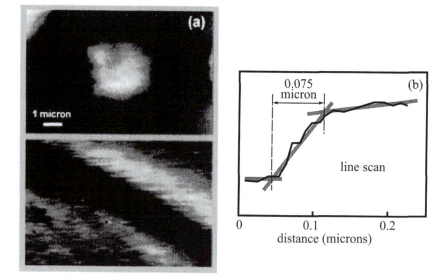

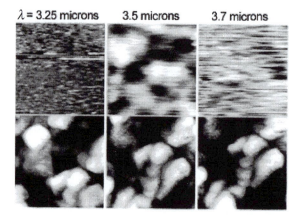

Fig. 4.14 (a), top: SNOM micrograph (taken with an FEL at $\lambda = 2.4$ microns) of platinum deposited on silicon, compared with a topographic (shear force) image ((a), bottom). The differences between the two images show that the SNOM features are not topographic artifacts. Data from: A. Cricenti, R. Generosi, P. Perfetti, J. M. Gilligan, N. H. Tolk, C. Coluzza and G. Margaritondo, *Appl. Phys. Lett.* **73**, 151 (1998). (b) Intensity line scan of a $\lambda = 2.4$ micron SNOM image detail. The scan shows that the lateral resolution is below 0.1 micron, much better than the diffraction limit.

Fig. 4.15 Example of microchemical analysis using a SNOM with an FEL. The top pictures are SNOM micrographs of diamond grains grown on a silicon substrate taken at three different wavelengths (3.25, 3.5 and 3.7 microns). The bottom pictures are the corresponding topographic microimages. The top-center image wavelength (3.5 microns) corresponds to the excitation of the C–H bond stretch vibration. Therefore, this image reveals the distribution of the chemical species corresponding to that bond. The two other SNOM pictures show no C–H related structure. Data from: A. Cricenti, R. Generosi, M. Luce, P. Perfetti, G. Margaritondo, D. Talley, J. S. Sanghera, I. D. Aggarwal , J. M. Gilligan and N. H. Tolk, *J. Microscopy (Oxford)* **202**, 446 (2001).

Inset Q: Overcoming the diffraction limit in microscopy

The absence of the diffraction limit in the SNOM geometry can be understood in terms of Werner Heisenberg's 'uncertainly principle'. This principle is a fundamental property of all particles, including photons. It states that we cannot *simultaneously* measure with infinite accuracy both the position of a particle and its momentum.

Specifically, when measuring at the same time the position y and the momentum p_y along the y-axis, the experimental uncertainties Δy and Δp_y must obey the law

$$\Delta y \Delta p_y \geq h ,\tag{Q1}$$

where h is again the Planck constant. Equation Q-1 implies in particular that if the accuracy in measuring y is perfect (Δy equals zero), then Δp_y for a simultaneous p_y-measurement is infinite—and viceversa.

Consider an optical magnifying system such as that of Fig. Q-1, based on the detection of photons scattered by the imaged object. Taking an image with the detector implies measuring both the positions of the scattered photons at the object plane and their scattering directions.

By measuring the scattering direction of a photon we also measure the two components p_x and p_y of its momentum. Equation Q-1 can thus be applied to the magnifying process, setting a limit for the accuracy in measuring y—and therefore in resolving image points close to each other. Specifically:

$$\Delta y \geq h/\Delta p_y .\tag{Q2}$$

The maximum possible Δp_y-value—corresponding to the minimum (best) Δy-value—cannot exceed p_y. In turn, since:

$$p = \sqrt{p_x^2 + p_y^2}\tag{Q3}$$

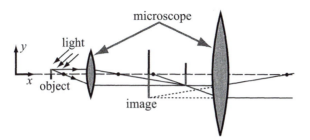

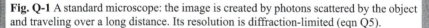

Fig. Q-1 A standard microscope: the image is created by photons scattered by the object and traveling over a long distance. Its resolution is diffraction-limited (eqn Q5).

the maximum p_y-value corresponds to a zero value for p_x, which gives $p_y = p$. On the other hand, the momentum p is linked to the wavelength of the particle by the second of De Broglie's equations:

$$p = h/\lambda \; ; \tag{Q4}$$

therefore

$$\Delta y \geq h/p = h/(h/\lambda) = \lambda \;. \tag{Q5}$$

This result sets the diffraction limit for the resolution along the y-direction, which is valid for a 'far-field' magnifying geometry like that of Fig. Q-1. The term 'far-field' means that the scattered waves travel over a long distance and are detected far away from the scattering object.

Assume, on the contrary, that the wave is an evanescent wave emanating from the tip of an optics fiber and interacting with an object very close to the fiber tip—as shown in Fig. Q-2. Along the forward x-direction the wave must be described by a progressively attenuated function like eqn F5 of Inset F. This means that the wavelength λ along the x-direction is replaced by $\lambda/(n + in_i)$—where n and n_i are the real and imaginary parts of the complex refractive index.

According to eqn Q2 (applied to the x-direction), this means that p_x is no longer real but a complex number with real and imaginary parts. Therefore, its square p_x^2 can be negative rather than positive, so that a zero p_x-value no longer corresponds to the maximum p_y-value in eqn Q3.

Thus, our previous derivation of the diffraction-limit (eqn Q5) is not valid for the geometry of Fig. Q-2. We cannot derive a minimum Δy-value linked to the wavelength as we did for Fig. Q-1, and the diffraction limit disappears.

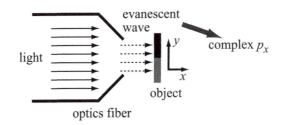

Fig. Q-2 In a near-field microscope, the forward wave is evanescent, thus the forward component of the photon momentum is complex and the diffraction limit is no longer valid.

4.3. X-ray FELs (X-FELs)

Can the FEL approach be extended to short-wavelength X-rays? This would be a very attractive possibility opening up many exciting new opportunities for research and technology. Unfortunately the obstacles are quite serious.

We have seen that the optical gain decreases when the wavelength decreases. Therefore, achieving lasing conditions is much more difficult when the spectral range changes from infrared light to X-rays.

We might think that a limited optical gain could still be compensated by an optical cavity. Unfortunately, no optical cavity exists for X-rays. Such a cavity, in fact, would require normal incidence mirrors—and we saw that normal-incidence reflection is very limited for X-rays. Reflection-enhancing techniques such as multilayer coating are not suitable in this case because of the high incident power.

The only possible strategy, therefore, is to compensate for the optical gain decrease at short wavelength by improving other FEL characteristics. The resulting gain must be sufficiently high to achieve laser action along a single wave path—with no optical cavity. This approach is known as 'super-radiance' for normal lasers. In the case of X-ray FELs, the key mechanism is called 'self-amplified spontaneous emission' or SASE.

The SASE approach optimizes the FEL operating parameters so that the initial spontaneous emission of the electron beam is amplified by stimulated emission reaching saturation before the wiggler end. This makes the external optical cavity unnecessary.

The SASE technique requires a high-quality electron accelerator of the LINAC type with high acceleration per unit length. The approach was successfully tested at progressively short wavelengths—see Fig. 4.16. However, the extension of this technique to X-rays still depends on the solution of some key technical problems.

We should note that each FEL source emits only one photon beam. Therefore, the operation and amortization costs cannot be easily spread over many simultaneous experiments. To avoid prohibitive costs for each experiment, clever schemes are proposed. They would specifically partition the accelerated electron beam among several independent wigglers thereby obtaining several FELs working in parallel.

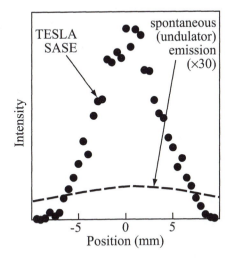

Fig. 4.16 The successful results of the first SASE tests at ultraviolet wavelengths obtained at the TESLA test facility of HASYLAB-DESY, Hamburg. Data derived from: J. Andruszkow *et al.*, *Phys. Rev. Lett.* **85**, 3825 (2000).

The implementation of X-FELs would have an exceedingly important impact on chemical, biological and medical research. The peak brightness would be increased by many orders of magnitude, as shown in Fig. 4.17. This brightness would be concentrated into pulses of extremely short duration—100 femtoseconds (one-tenth of a millionth of a millionth of a second) or even less. In other words, an X-FEL would confine extremely high electromagnetic power in a short time, opening up unexplored domains of chemistry.

We should note, however, that these performances are complementary to those of a storage ring source, whose peak power is much lower but whose effective pulse length is much longer. Therefore, the comparison between normal synchrotrons and X-FELs is similar to that between continuous (DC) and pulsed conventional lasers. Both types of sources are important, their roles are largely complementary, and one type cannot replace the other.

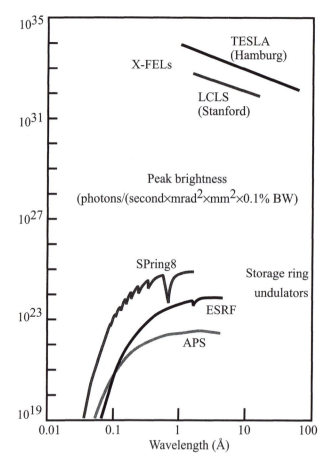

Fig. 4.17 The predicted peak brightness of future X-FELs is many orders of magnitude larger than the best present undulators. Note, however, that the difference is much smaller for the average brightness, and that the two types of sources will play complementary roles.

Among the many new experiments that might be possible with X-FELs, we should mention one-shot imaging and one-shot crystallography. The basic idea in both cases is to obtain data by using a single, extremely energetic pulse of photons. The pulse would destroy the sample, but we could obtain results so fast that they would correspond to an almost undamaged sample. The corresponding time scale would range from one-hundredth to one-tenth of a millionth of a millionth of a second, in the same general range as the projected X-FEL pulse duration.

Why one-shot crystallography? We saw that protein crystallography requires crystals of reasonable size and quality. In many cases these requirements cannot be met. One-shot 'crystallography' could analyze individual molecules without requiring crystals.

Are these just dreams or future reality? The forthcoming years will give us the answer. One point, however, is already clear: FELs are steadily moving from large (infrared) wavelengths to shorter wavelengths. Independent of the full development of SASE for X-rays, FELs are already becoming attractive sources for the ultraviolet range.

5. Future directions

Synchrotron-based activities have expanded for more than three decades. This evolution is not slowing down but accelerating.

We have already discussed several important aspects of the expansion such as the shift from physics to the life sciences and the progress towards X-FELs. There are many other important aspects. In this chapter we discuss some of them, in three general areas: new sources, new applications and new management approaches.

5.1. New sources

FELs and in particular X-FELs are very exciting new sources. However, they are not the only innovating concept in this area.

We should specifically note compact X-ray sources based on the 'Compton backscattering' effect. This phenomenon is illustrated in Fig. 5.1: an infrared laser beam and a high-energy electron beam collide head-on. The laser beam takes energy from the electrons while it is backscattered, increasing the energy of its photons and decreasing their wavelength.

This transforms infrared laser light into collimated X-rays. The phenomenon is easily understood with arguments similar to those used in Section 1.1.2.3 for the undulator emission.

Assume that the infrared laser wavelength is λ_I. Imagine the laser beam as seen by the electrons moving with speed close to c—corresponding to a large γ-factor. The electrons 'see' the source (laser) moving at high speed. As a consequence, a Doppler shift modifies the observed wavelength.

This (see the derivation of eqn 1.4) changes λ_I into $\approx \lambda_I/(2\gamma)$. Subject to the Doppler-shifted wave, the electrons undulate in the transverse direction and emit a wave of the same wavelength $\approx \lambda_I/(2\gamma)$. This is the Compton-backscattered wave.

Note that the backscattered wavelength is $\approx \lambda_I/(2\gamma)$ when measured in the reference frame *of the electrons*. When observed in the *laboratory* frame, this wave is emitted by a moving source—the electrons—and therefore again subject to a Doppler shift. The Doppler factor is once again $\approx 2\gamma$. Therefore, the doubly Doppler-shifted backscattered wavelength is

$$\lambda_X \approx [\lambda_I/(2\gamma)]/(2\gamma) = \lambda_I/(4\gamma^2) . \tag{5.1}$$

The large γ-factor corresponding to a large $4\gamma^2$-value explains the wavelength conversion from infrared light to X-rays. This approach can be implemented with a compact electron accelerator rather than with a large storage ring. A Compton-backscattering source can thus be quite small and suitable, for example, for a radiological facility in a hospital.

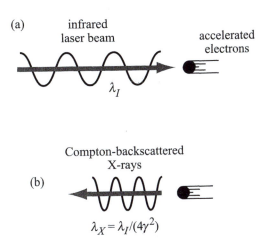

Fig. 5.1 Compton backscattering: an infrared laser beam collides head-on with electrons moving at nearly the speed of light *c*. After a double Doppler shift, the backscattered photons have much shorter wavelengths: the infrared laser beam becomes an I-ray beam.

Compton backscattering sources do not reach top performances as far as geometry and coherence are concerned, and are inferior to advanced undulators or FELs for applications such as spectroscopy or microscopy. However, small size, limited cost and flexible operation makes them potentially attractive for non-conventional radiology.

When brightness, collimation and coherence are prime objectives, standard storage rings equipped with undulators are still excellent sources. The storage ring technology has not yet reached its performance limits. Advanced designs may improve brightness and coherence by one or two orders of magnitude with respect to the best present sources.

The effectiveness of storage rings can be further improved by replacing the standard injection of electrons in the ring with a 'top-off' procedure. In this case electrons are regularly injected at short time intervals rather than once every several hours—and the beam current is kept almost constant.

Other approaches are based on entirely new concepts. Particularly interesting are the *energy-recovery accelerators*. Such sources use each injected electron bunch only once to emit light, whereas a storage ring uses the same circulating electrons over and over for a long time.

The one-emission-per-bunch approach offers several advantages. Since the bunch does not circulate many times, the accelerator does not have to be designed for this function. It can be optimized instead for preserving the electron beam geometry delivered by the injector. Very high brightness is thus possible compared with storage rings.

However, by using each electron bunch only once the source would consume too much power and its operating costs would be prohibitive. The solution is provided by the energy-recovery scheme.

Each injected electron bunch is first accelerated, then used to emit light, then decelerated giving energy back to the accelerator system, and finally dumped. The recovered energy substantially decreases the effective power consumption.

Among the energy-recovery designs, we should note the MARS (Multi-turn Accelerator Recuperator Source) project by Gennadi Kulipanov and his Novosibirsk team—see Fig. 5.2. In the first (energy-taking) phase, the electron bunch moves along a series of different paths corresponding to increasing electron energies. Such paths correspond instead to decreasing energies in the energy-recuperation phase. Each path is equipped with an undulator for the emission of synchrotron light.

This schemes offers many advantages—specifically the possibility of using very long undulators. Why long undulators? We saw in Section 1.3.1 (eqn 1.11) that

$$\frac{\Delta\lambda_L}{\lambda_L} = \frac{1}{nN},$$

which means that the relative bandwidth $\Delta\lambda_L/\lambda_L$ decreases as the number of undulator periods N increases. A long undulator can accommodate a large number of periods and produce a small bandwidth $\Delta\lambda_L$.

Could $\Delta\lambda_L$ become so small that monochromators are no longer required? This would be a very attractive possibility since all monochromators have limited throughput and strongly reduce the light delivered to the experimental chambers.

Unfortunately, the decrease in $\Delta\lambda_L$ by increasing N is limited by the so-called 'energy spread' of the electrons (already mentioned in Section 4.1). This spread corresponds to the small differences in energy between different circulating electrons.

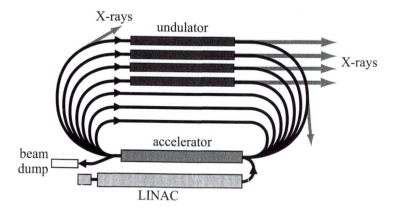

Fig. 5.2 Simplified scheme of MARS (Multi-turn Accelerator-Recuperator Source). In the accelerating phase, the electrons pass through the LINAC and then through the accelerator. At each passage their energy is increased and they follow a longer trajectory. The sequence is reversed in the energy-recuperation phase and ends in the beam dump. X-rays are produced by very long undulators and also by bending magnets. Details on the MARS project can be found in: G. N. Kulipanov, A. N. Skrinsky and A. Vinokurov, *J. Synchrotron Rad.* **5**, 176 (1998).

How does the electron energy spread affect the wavelength bandwidth? Consider eqn 1.4:

$$\lambda_L \approx \frac{L}{2\gamma^2}.$$

This equation implies that the γ-factor determines the emitted undulator wavelength. The γ-factor corresponds to the electron energy. The electron energy spread translates into a γ-factor spread, which in turn causes a wavelength spread.

$\Delta\lambda_L$ cannot become smaller than this wavelength spread. When this limit is reached, an additional increase in N has no effect on the wavelength bandwidth. In practice, the decrease in $\Delta\lambda_L$ saturates for $N = 100$–200. The corresponding $\Delta\lambda_L/\lambda_L$ is still too large for most applications. In conclusion, monochromators cannot be eliminated by using long undulators with many periods.

Could this negative result be reversed by reducing the electron energy spread? For storage rings, the answer is negative. The energy spread is primarily due to synchrotron light emission, i.e. to the very scope of a synchrotron source. In fact, the emission of photons decreases the energy of the emitting electrons with respect to other circulating electrons.

The MARS approach proposes a clever solution to this problem. As seen in Fig. 5.2, an electron reaches an undulator after emitting synchrotron light at lower energies (in previous undulators). Synchrotron light emission (see Chapter 1) increases very rapidly with the electron energy. Thus, in this scheme the emission and the consequent energy spread are reduced with respect to the storage rings, which work with constant average electron energy.

A MARS source could thus accommodate very long undulators with large N-values to decrease $\Delta\lambda_L$ beyond the storage-ring limit—and eliminate the monochromators for many applications. Practical tests must of course demonstrate the feasibility of MARS and of other energy-recuperation schemes.

5.2. New applications

In addition to the examples of the previous sections, synchrotron sources find new and often unexpected applications in novel areas. The most relevant techniques for such novel applications belong to chemical and structural analysis—with particular emphasis on non-destructive tests.

We have already mentioned, for example, that synchrotron-based tests are increasingly used in environmental studies. A similar trend is observed in geophysics and geochemistry.

Non-destructive tests are also very important for archaeological objects. Synchrotron-based archaeometry is in fact a rapidly expanding domain, providing important new types of data for archaeological specimens. In the future, we foresee similar approaches applied to the certification of art objects and to the analysis of historical specimens.

Synchrotron-based tests are already successfully used in some interesting history studies. For example, X-ray fluorescence with synchrotron radiation was recently used to study a hair specimen belonging to Ludwig Van Beethoven [D. C. Mancini *et al.*, private communication]. The results revealed an anomalous lead content suggesting long-term lead poisoning and opening up interesting historical questions.

5.3. Novel management techniques

The management of a synchrotron facility has one primary objective: meeting the needs of users in producing good science. Since the clientele is changing, the management techniques must evolve. The present approaches were primarily developed for physics and do not always fit new users in biology, medicine and several other areas.

Consider the projected cost of synchrotron radiological tests. The construction of a large facility costs US$ 100–200 million (this applies to a prototype source and could be lowered by large-scale construction). Synchrotrons have a very long lifetime, so the amortization costs are reasonably limited, US$ 5–10 million per year.

A conservative estimate of the operation cost is US$ 20–30 million per year. The total amortization and operation cost would then be US$ 25–40 million per year.

Such a facility could support up to 80-100 beamlines working 5000 hours per year. Assuming conservatively only 30 radiological stations, the facility would deliver 30 × 5000 = 150000 beamtime hours per year. With one radiological examination per station per hour, the instrumentation cost of each examination would be (US$ 25–40 million) divided by 150,000, which gives US$ 170–270. The total examination cost must of course take into account the medical personnel and the clinical infrastructure.

Are these cost too elevated? The answer depends on the benefits for the patients and for society. Note that the above values are upper estimates under very conservative assumptions. Realistic values could be a factor of two or three lower or even less. Furthermore, the evaluations could radically decrease for a small source (e.g. a Compton backscattering source).

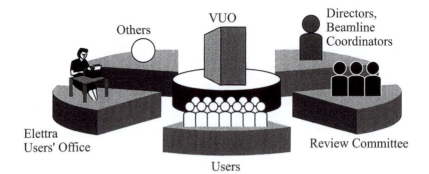

Fig. 5.3 Schematic illustrating the Virtual Users's Office (VUO) software package developed at Elettra by R. Pugliese *et al.* The VUO can handle most of the user-facility interface functions through the Internet in a fully interactive way.

Financial issues, in summary, do not appear to be an impassable barrier. Technical problems are difficult but not impossible to solve. Thus, synchrotron radiology facilities could become a reality and would require completely different management techniques with respect to synchrotron research facilities.

Management changes, on the other hand, are also necessary for standard synchrotron research facilities. Access to synchrotron beamtime has so far been granted (based on merit) for specific single experiments. This was until now a fair and effective procedure.

The needs of customers from some new domains, however, cannot be met with this approach. Protein crystallography, for example, consists of short experiments within broad programs, with fast beamtime access at largely unpredictable dates.

This requires new procedures such as 'block' beamtime allocation to programs rather than to single experiments. Similarly, industrial use often implies fast-track procedures (with beamtime payment for proprietary research).

The Internet has a strong impact on synchrotron management; for example, the beamtime allocation is increasingly handled on-line. The most advanced approach is the so-called 'Virtual Users' Office', a software package developed and successfully tested by the Elettra facility in Trieste—see Fig. 5.3.

5.4. A few final words

This author has worked with synchrotrons sources since 1971, starting with a real synchrotron in Frascati (within the pioneering PULS project led by Gianfranco Chiarotti and Franco Bassani). His first dedicated storage ring—Tantalus in Wisconsin, designed and directed by Ednor M. Rowe—had a diameter of a few meters and an amazingly small cost compared with recent facilities. It is impossible not to be excited by the sight of the new gigantic facilities—with hundreds of meters of circumference, tens of beamlines, thousands of users and a huge flow of scientific and technological results.

These monsters—however wonderful for synchrotron radiation veterans—can be intimidating for prospective new users. This is regrettable: facilities that are so powerful and useful should be inviting rather than intimidating.

The best way to eliminate the intimidation is to understand what a synchrotron is and how it can be very useful. This learning process is facilitated, in this author's opinion, if the presentation is in simple terms and free from formalism. This was the objective of our book. We sincerely hope that it was reached and that the book will stimulate the interest and creativity of many new synchrotron users.

Index